CENTRIFUGAL PUMPS

BY

J. W. CAMERON, A.M.I.Mech.E.

WITH EIGHTY-FOUR ILLUSTRATIONS AND DIAGRAMS

LONDON

SCOTT, GREENWOOD & SON

8 BROADWAY, LUDGATE, E.C. 4

1921

PREFACE.

By reason of the development of the electric motor and steam turbine, the general trend of present day practice is to substitute rotary for reciprocating machinery. In fact, the time is not far distant when the centrifugal or turbine pump will entirely displace the reciprocating type of pump.

In writing this book the author assumes that the reader has a knowledge of the elementary principles of hydraulics. The examples on the design of centrifugal pumps illustrate the method used by the author in fixing up the principal dimensions of this type of pump and may serve as a guide to the young designer.

It is hoped that this volume may prove useful to Engineers, Draughtsmen, and Students.

For the use of many illustrations appearing in the pages of this work, the writer would acknowledge his great obligation to the following engineering firms : The American Well Works, U.S.A. ; Sulzer Bros., Switzerland ; Hayward Tyler & Co., Ltd., London.

J. W. CAMERON.

LUTON, *June,* 1921.

CONTENTS.

CHAPTER VII.

CHAPTER VIII.

CHAPTER IX.

CHAPTER X.

CHAPTER XI.

CHAPTER XII.

NOTATION.

The notation adopted throughout this book is in accordance with the following list :—

a	$=$ Constant $1 - m^2$, also areas of passages.
b_1, b_2	$=$ Width of runner at inlet and outlet (feet).
d, d_1, d_2, d_3	$=$ Diameters.
f	$=$ Stress (lbs. per square inch).
g	$=$ Acceleration due to gravity, 32·2 ft. per sec. per sec.
h_s	$=$ Distance from centre of pump to surface of water in suction tank.
h_d	$=$ Distance from centre of pump to discharge outlet.
l	$=$ Length of pipe (in feet).
$l_1, l_2,$ and etc.	$=$ Losses of head.
m	$=$ Ratio $\dfrac{r_1}{r_2}$.
n	$=$ Revolutions per min.
p_1 and p_2	$=$ Pitch of runner vanes.
p, p_s, p_d	$=$ Pressure (lbs. per square foot).
q	$=$ Ratio $\dfrac{w_1}{w_2}$.
r_1 and r_2	$=$ Radius of runner at inlet and outlet.
r_3, r_4	$=$ Radius.
t_1, t_2	$=$ Thickness of vanes.
u	$=$ Peripheral velocities.
v	$=$ Absolute velocities.
w	$=$ Relative velocities.
x	$=$ Ratio $\dfrac{w_2 \sin \beta_2}{u_2}$.
D	$=$ Diameter of runner.
F	$=$ Area.
G	$=$ Density of fluid.

H	$=$	Head in feet.
I	$=$	Moment of inertia.
M	$=$	Bending moment.
N	$=$	Horse-power.
Q	$=$	Quantity.
T	$=$	Twisting moment, end thrust in lbs.
W	$=$	Weight (lbs.).
a	$=$	Angle between absolute velocity and tangent at any radius.
a_2, a_3, a_4	$=$	Guide vane angles.
β_1, β_2	$=$	Runner vane angles.
ζ, ϕ, ψ	$=$	Coefficients.
η_h	$=$	Hydraulic efficiency.
η_o	$=$	Overall efficiency of pump only.
μ	$=$	Coefficient of friction.
ω	$=$	Angular velocity radians per second.

ERRATA.

Page 19:— $\qquad q = \dfrac{1}{m}\dfrac{l_2 \sin \beta_2}{l_1 \sin \beta_1}$, where $\dfrac{l_2}{l_1} = \dfrac{1}{1 \cdot 75}$.

should read

$$q = \dfrac{1}{m}\dfrac{b_2 \sin \beta_2}{b_1 \sin \beta_1}, \text{ where } \dfrac{b_2}{b_1} = \dfrac{1}{1 \cdot 75}.$$

Page 33 at foot should read

$$\frac{p_d - p_1}{G} = \frac{u_2^2 - u_1^2}{2g} + \frac{w_1^2 - w_2^2}{2g} + \frac{2(v_2\,_c \cos a_2 - v_c^2)}{2g} - \frac{v_d^2}{2g} + \frac{v_c^2}{2g} - l_1 - l_2 - l_3 - l_7.$$

Page 34, second formula, should read

$$H_m = \frac{p_d - p_1}{2g} - \frac{v_1^2}{2g}$$
$$= \frac{u_2^2 - u_1^2}{2g} + \frac{w_1^2 - w_2^2}{2g} + \frac{2v_2 v_c \cos a_2}{2g} + \frac{v_c^2}{2g} - \frac{v_d^2}{2g} - \frac{v_1^2}{2g} - l_1 - l_2 - l_3 - l_7,$$

Page 90, Fig. 48A' should be reversed.

Centrifugal Pumps.

CHAPTER I.

THE turbine or centrifugal pump is an apparatus for utilising the mechanical energy of a· prime mover, in order to give to water, or any other fluid, potential energy.

It is, in fact, a reversed radial inward flow reaction turbine, which is supplied with power from an external source, parting with it to the fluid passing through instead of supplying power to an outside source.

For many years a crude form of turbine pump called a centrifugal pump has been used for the elevation of large volumes of water to moderate heights, giving satisfactory results under these conditions.

The credit for the design of the first centrifugal pump, although of a crude type, seems to be due to the great mathematician, Euler; it was not until about the year 1850 that the centrifugal pump came into commercial use.

The first improvement in this type of pump was to the shape of the runner vanes, whereby greater efficiency was obtained over the radial vane, with the type of casing in vogue in those days.

In the early days, these pumps were driven by reciprocating steam engines, either with the crank shaft coupled direct to the pump shaft, or else the pump was driven from the engine through the medium of a belt or ropes.

The first form of drive was the most reliable, but on account of the low speed of the engines then obtainable, it was impossible to raise water to any but comparatively low heads with reasonable efficiency.

I

This has been changed by the placing on the market of
reliable high speed motors, such as electric motors and steam
turbines which can be coupled direct to this type of pump,

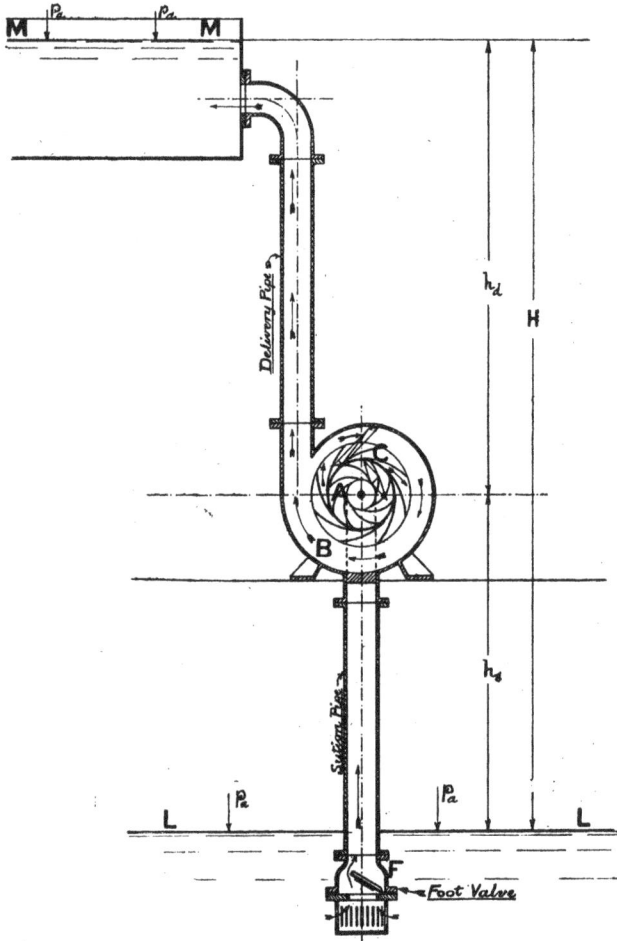

Fig. 1.—Centrifugal or turbine pump installation.

resulting in a great improvement, both in the design and
efficiency of the centrifrugal pump.

Centrifugal or turbine pumps can now be employed in

elevating fluids to heights which previously could only be dealt
with by reciprocating pumps.

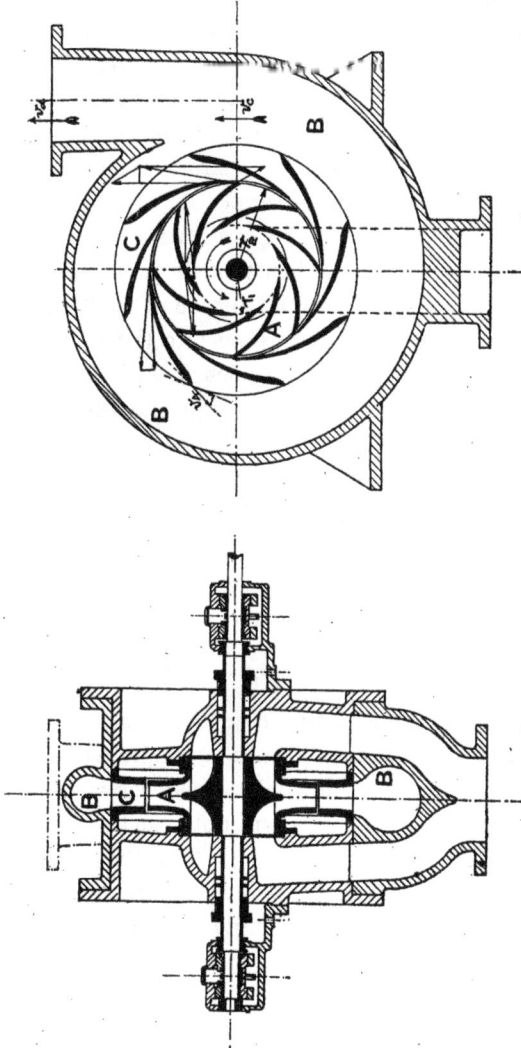

FIG. 2.—Turbine pump or centrifugal pump with guide vanes.

The illustration in Fig. 1 represents a typical contrifugal
or turbine installation, and Fig. 2 represents sections of the
pump only.

The pump consists of a disc or runner A, provided with a number of vanes revolving within a casing B, containing a ring of guide vanes C, often called the diffuser.

As only pumps dealing with water and similar liquids are under consideration, it will be understood that where the term fluid is used liquids of this description are referred to.

Therefore, in the calculations which follow it is always assumed that the pump is fully charged with such fluid from the foot-valve F to the top of runner A.

Under usual conditions, the pressure of the suction inlet or eye of the runner is less than that of the atmosphere.

The pressure p_a, acting on the surface of the fluid LL, causes the fluid to rise up the suction pipe into the runner (in which its pressure is increased), a continuous flow of fluid through the pump resulting. In passing through the runner energy is given to the fluid; the difference between the level LL and level MM is the measure of that part of this energy which is usefully applied, and for each unit weight of fluid is equal to the vertical height H, called the static head.

The water particles flowing from the suction pipe into the runner at the radius r_1, Fig. 2, have a velocity equal to v_1, and the kinetic energy per unit weight of fluid is therefore $\dfrac{v_1{}^2}{2g}$.

In order that there shall be no shock at entry, the inlet edges of the vanes must make such an angle with tangent to the circle of radius r_1, at the point where the vanes start, that the fluid is sliced by the vanes and not struck.

To meet this condition, the angle that the tangent to the first element of the vane must be inclined, with the tangent to the circle of radius r_1, is such an angle B_1 that tan B_1 is equal to the absolute velocity v_1, divided by the peripheral velocity u_1 at radius r_1.

This is shown by the parallelogram of velocities at entrance A in Fig. 3, w_1 being the relative velocity of the fluid at inlet, that is, the velocity of flow of the fluid along the vane.

The fluid then flows through the runner passages (formed

by two consecutive vanes from A to B) from entry to exit of runner, leaving it with an absolute velocity v_2.

The absolute velocity v_2 is the diagonal of the parallelogram of velocities shown at the point B, Fig. 3, formed by the peripheral velocity u_2 and the relative velocity w_2 as sides.

On leaving the runner the fluid enters the diffuser provided

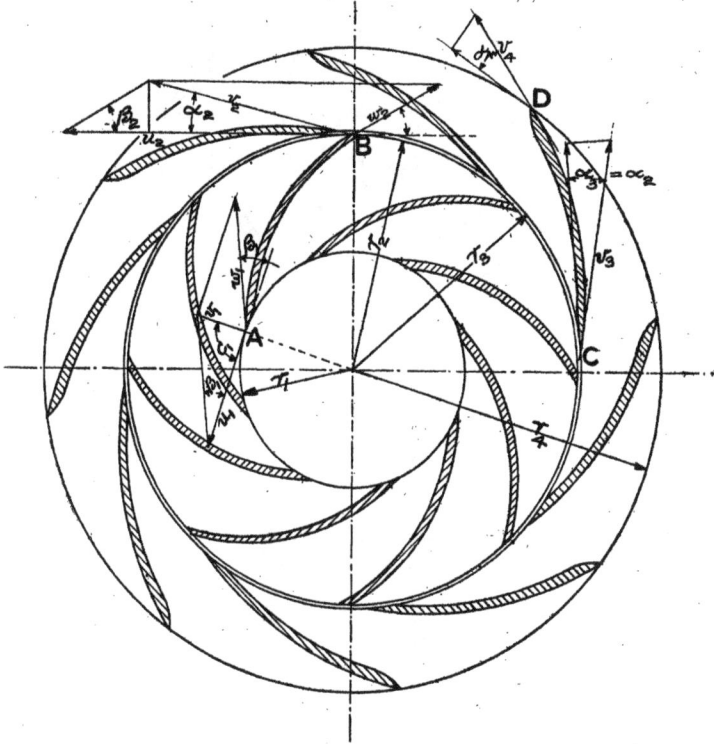

FIG. 3.—Section of runner vanes and guide vanes.

with guide vanes, so designed that the fluid enters without shock, and during its flow along the passages formed by two consecutive vanes, its velocity is gradually reduced from v_3 at C, Fig. 3, to v_4 at D, Fig. 3, and consequently this reduction of kinetic energy appears as an increase in pressure energy.

From the outlet of the diffuser at D, Fig. 3, the fluid flows into and around the casing of continually increasing cross section B, Fig 2, to the discharge pipe, a further increase in

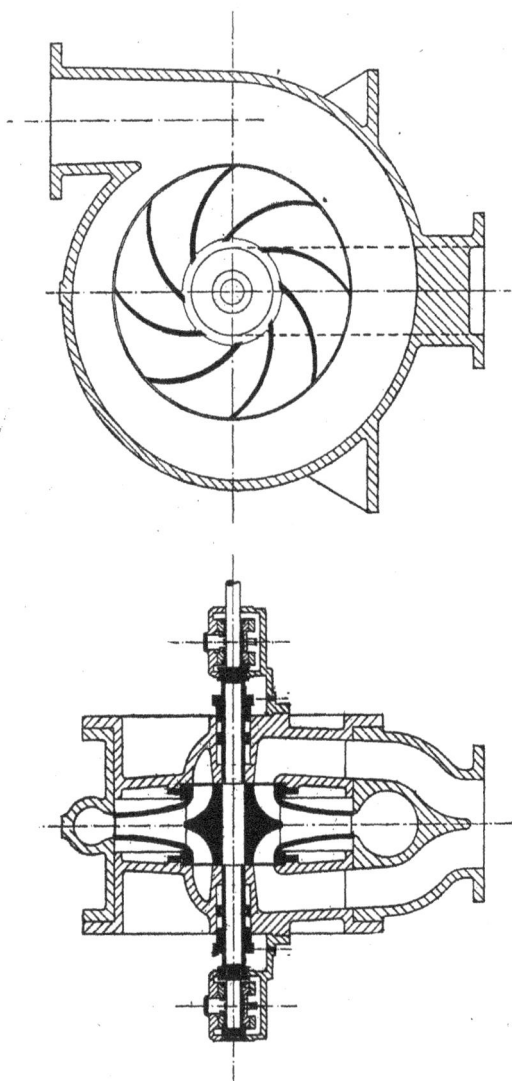

FIG. 4.—Centrifugal pump with spiral casing.

pressure energy taking place if the casing B is properly designed.

This type of pump, as represented by Figs. 1 and 2, generally represents the type adopted to obtain high efficiencies, or where the least expenditure of power is of vital importance.

Where first cost and not efficiency is the principal consideration, the guide vanes and diffuser are done away with, the fluid being discharged directly from the runner into the spiral casing.

This type of pump is represented by Fig. 4, and is a favourite type with some pump builders. One of the advantages they claim for it is simplicity, while they hold that it is specially advantageous in cases where the water is taken from rivers, lakes, etc., and the suction pipe inlet cannot be screened sufficiently to prevent the entrance of solid matter carried by the water in suspension.

In the latter case this type of pump is popular on account of there being no guide or diffusion vanes to clog up or renew owing to breakage.

However, in cases where the fluid is efficiently screened, and settling chambers are used to enable heavy solid particles to settle, then, for permanent installations, the type of pump with guide vanes (Figs. 2 and 3) is the most efficient and economical.

Fig. 5 represents a cheaper and lighter type of pump than either of the types described above. This design is the least efficient of all and should only be used for comparatively low heads, and temporary work, such as the unwatering of foundations, etc. This type is still built by some builders, especially for circulating the cooling water in surface condensers on shipboard, the pump being direct-coupled to a high speed steam engine.

The advantage of centrifugal or turbine pumps over other types is simplicity in both construction and working.

They have no valves, therefore there is no interruption in the water column, with the accompanying inertia forces due to the starting and stopping of the liquid column.

Should the delivery stop while the pump is working, owing to the closing of a valve or an obstruction in the delivery pipe, the pressure in the delivery main and pump chamber increases only by a small amount above the normal head.

In some designs, this pressure, when the pump is delivering no water, is even less than that at normal load. It de-

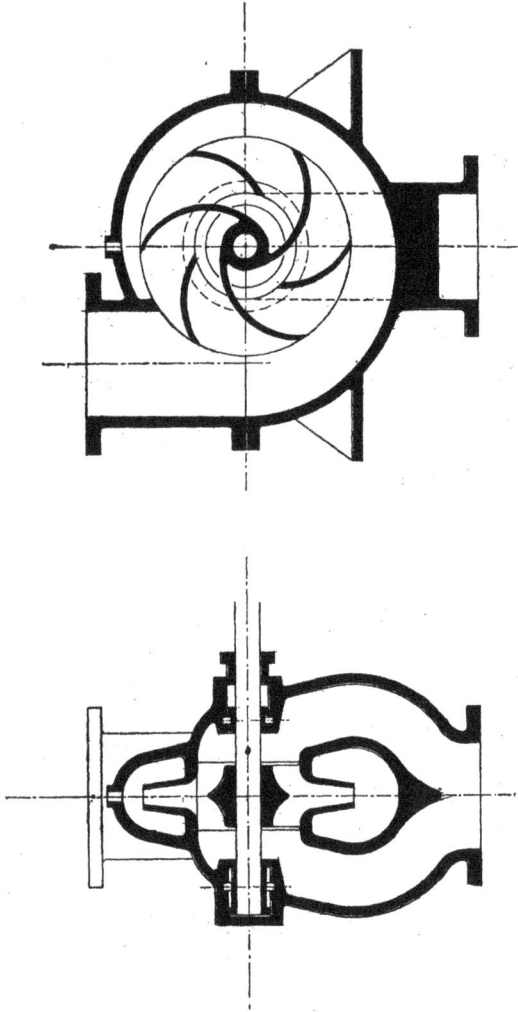

FIG. 5.—Early type of centrifugal pump.

pends solely on the speed of rotation whilst the power absorbed by the pump is less than when the pump is delivering
its normal quantity.

Therefore, relief valves that are necessary to safeguard reciprocating pumps and to prevent breakage under these conditions, are unnecessary in the case of centrifugal pumps.

Practically the only attention required with a centrifugal pump is to see that the bearings have the necessary supply of oil in their reservoirs, and occasional packing of the stuffing boxes.

The space required for the centrifugal pump is very much less than that required for any other type for the same output ; this means lighter and cheaper foundations.

As regards first cost, the advantage is with the centrifugal pump, the cost of the centrifugal pump and electric motor in some cases being one-fifth the cost of a reciprocating pump of the same output with its electric motor.

Also the speed of rotation of this type of pump corresponds to the most advantageous speed of electric motor ; this enables pump to be direct-coupled to the motor, doing away with all forms of gear.

Providing the quantity of water to be dealt with is not too small, when compared with head per stage of pump, the efficiency of the centrifugal or turbine pump compares very favourably with the best types of reciprocating pump, and there is no duty for which a centrifugal pump cannot be utilised.

CHAPTER II.

THE THEORY OF THE CENTRIFUGAL PUMP.

IN the following theory, for simplicity, the assumption will be made that the axis of the pump is vertical, in order that particles at equal distances from the centre of the runner may have equal pressures and velocities.

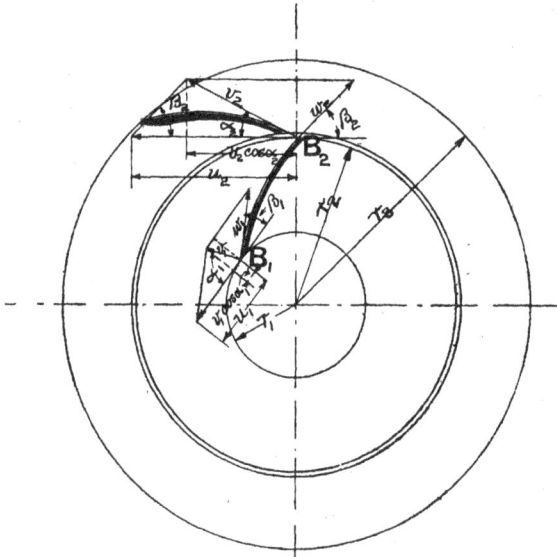

FIG. 6.—Diagram of runner and guide vane.

Also that the velocity of flow through a channel is the same at all points on a cross section.

Referring to Figs. 6 and 7, B_1B_2 represents the vane of a centrifugal pump runner, the inner and outer radii being r_1 and r_2 respectively.

A particle of water enters the runner at B_1, in the direction

(10)

B_1V_1, with an absolute velocity v_1, making the angle a_1 with the tangent to B_1 at radius r_1, B_1U_1 is the peripheral velocity of the runner at radius r_1, then the relative velocity w_1 along the vane at entry B_1W_1 is the vector sum of v_1 and $- u_1$.

The absolute velocity v_1 is then the diagonal of a parallelogram whose sides are u_1 and w_1, the angle β_1 that w_1 makes with u_1 is the angle of the vane at entry (Figs. 6 and 7). If no guide vanes are used at entry, $\dfrac{v_1}{u_1}$ is made $= \tan \beta_1$.

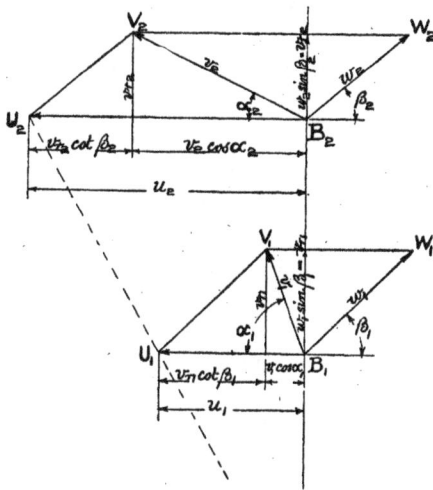

FIG. 7.—Entrance and outlet diagrams.

The water particles flow along the vane, leaving it at B_2 with an absolute velocity v_2 in the direction B_2V_2, making the angle a_2 with the tangent of the circle of radius r_2 at B_2.

This is the direction along which the water particles are flowing the instant they leave the runner, and if guide vanes are used a_2 must be the angle that the first element of the guide vane should make with the tangent to the circumference of the runner at B_2, in order that the discharge from the runner into the guides may take place without shock.

The absolute velocity v_2 at exit from the runner is the resultant of the peripheral velocity u_2 in the direction B_2U_2

and w_2 the relative velocity along the vane in the direction B_2W_2, and is represented by diagonal B_2V_2 of the parallelogram of velocities $B_2U_2V_2W_2$.

The angle β_2 between the direction of u_2 and w_2 is the angle that the last vane element must make with the tangent to the circumference at B_2.

The absolute velocities v_1 and v_2 can each be resolved into two components at right angles to one another, one being in the direction of u_1 and u_2, the peripheral velocities, and other components being at right angles to u_1 and u_2 or along the radius.

The first or horizontal components are termed the velocities of whirl at entry and exit, and are respectively $v_1 \cos a_1$ and $v_2 \cos a_2$.

The components along the radii at B_1 and B_2 are respectively $v_1 \sin a_1$ and $v_2 \sin a_2$.

From the velocity diagrams at entry and discharge it will be obvious that the radial components are

$$v_{r1} = v_1 \sin a_1 = w_1 \sin \beta_1$$
$$\text{and } v_{r2} = v_2 \sin a_2 = w_2 \sin \beta_2.$$

Also that the velocities of whirl are

$$v_1 \cos a_1 = u_1 - v_{r1} \cot \beta_1$$
$$v_2 \cos a_2 = u_2 - v_{r2} \cot \beta_2.$$

Energy Imparted to the Fluid by the Runner.

Let v_1 = absolute velocity of the fluid at entry to runner.

$v_2 =$,,   ,,   ,, ,,   ,, ,, exit of   ,,

$w_1 =$ relative   ,,   ,, ,,   ,, ,, entry to   ,,

$w_2 =$,,   ,,   ,, ,,   ,, ,, exit of   ,,

$u_1 =$ peripheral ,,   ,,   ,,   ,, ,, entry to   ,,

$u_2 =$,,   ,,   ,, ,,   ,, ,, exit of   ,,

Q = quantity of fluid pumped in one second in cub. ft.

G = density of fluid.

The velocity of whirl, or the tangential component of v_1, is $v_1 \cos \iota v$, therefore the momentum of the fluid at inlet is

$$\frac{GQ}{g}(v_1 \cos a_1).$$

The momentum of the fluid at outlet of runner

$$= \frac{GQ}{g}(v_2 \cos a_2).$$

The change of momentum of fluid during its passage through the runner

$$= \frac{GQ}{g}(v_2 \cos a_2 - v_1 \cos a_1).$$

Change of moment of momentum

$$= \frac{GQ}{g}(r_2 v_2 \cos a_2 - r_1 v_1 \cos a_1).$$

If the runner is revolving with an angular velocity ω, the work done on the fluid per second during its passage through the runner

$$= E = \frac{GQ}{g}\omega(r_2 v_2 \cos a_2 - r_1 v_1 \cos a_1).$$

If r_1 and r_2 be the radii of the runner at inlet and outlet respectively, then $u_1 = r_1\omega$ and $u_2 = r_2\omega$.

Then $\quad E = \dfrac{GQ}{g}(u_2 v_2 \cos a_2 - u_1 v_1 \cos a_1).$

If H_t represent the maximum theoretical head, then this is equal to the energy imparted by the runner to every lb. of fluid passing through the pump, therefore

$$H_t = \frac{I}{g}(u_2 v_2 \cos a_2 - u_1 v_1 \cos a_1), \qquad . \qquad . \quad (1)$$

the energy per lb. of fluid existing partly as pressure and partly as kinetic energy.

This would also be the height that each lb. of fluid would be raised by the pump if the whole of the kinetic energy could be utilised and there were no shock losses and friction losses due to flow through pump.*

When no guides are used at entry to runner the absolute velocity at entry is assumed to be radial, or the fluid has no velocity of whirl, the entrance diagram will be as shown in Fig. 8.

In this case $v_1 \cos a_1 = 0$, since $a_1 = 90°$

and $$\frac{v_1}{u_1} = \tan \beta_1.$$

Under this condition equation (1) is simplified, becoming

$$H_t = \frac{1}{g} u_2 v_2 \cos a_2 . \qquad . \qquad . \qquad . \qquad (2)$$

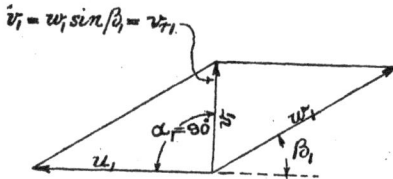

$\dot{v}_i = w_i \sin \beta_i = v_{r_1}$

Fig. 8.—Inlet velocity diagram.

Equations (1) and (2) may be written in terms of the peripheral velocities, the radial velocities, and the vane angles β_1 and β_2.

Then

$$H_t = \frac{1}{g}\{u_2(u_2 - v_{r_2} \cot \beta_2) - u_1(u_1 - v_{r_1} \cot \beta_1)\} \quad . \quad (3)$$

or when $a_1 = 90°$

$$H_t = \frac{1}{g} u_2(u_2 - v_{r_2} \cot \beta_2) \qquad . \qquad . \qquad . \qquad . \qquad (4)$$

$$= \frac{1}{g}(u_2^2 - u_2 v_{r_2} \cot \beta_2).$$

* From the parallelograms of velocity it will be easily seen that

$$\cos a_1 = \frac{u_1^2 + v_1^2 - w_1^2}{2u_1 v_1}$$

$$\cos a_2 = \frac{u_2^2 + v_2^2 - w_2^2}{2u_2 v_2}.$$

If Q_t = quantity passing through the runner in cubic feet per second,

Q = quantity flowing through the discharge pipe,

Then $Q_t - Q$ = the amount of fluid short circuited and lost by leakage, this loss depending on the clearances of the joints at eye of pump and the pressure differences.

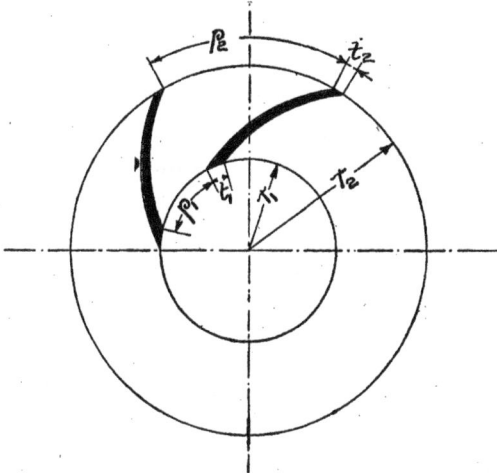

FIG. 9.—Diagram of runner vanes.

The runner must, therefore, be designed to deal with the quantity Q_t.

If F_2 = area of runner at outlet radius r_2

F_1 = ,, ,, ,, ,, inlet ,, r_1

$$F_2 = 2\pi r_2 b_2 \frac{p_2}{p_2 + t_2}$$

$$F_1 = 2\pi r_1 b_1 \frac{p_1}{p_1 + t_1}$$

Substituting these values in equation (1) we obtain the value of H_t in terms of the velocities

$$H_t = \frac{u_2^2 - u_1^2}{2g} + \frac{v_2^2 - v_1^2}{2g} + \frac{w_1^2 - w_2^2}{2g}.$$

This shows that H_t is equal to the sum of the increase in centrifugal force per lb. of fluid, the increase in kinetic energy, and the increase in pressure energy per lb. of fluid, due to flow in a runner passage of varying cross section.

This expression forms a useful check on the calculation of H_t after the velocity diagram has been drawn, and also shows the effect of altering any of the terms.

where b_1 and b_2 are the widths of the runner at inlet and outlet,
t_1 and t_2 are the thicknesses of the vanes measured along the
circumferences of circles of radius r_1 and r_2, p_1 and p_2 are the
distances between two consecutive vanes at inlet and outlet
respectively (see Fig. 9).

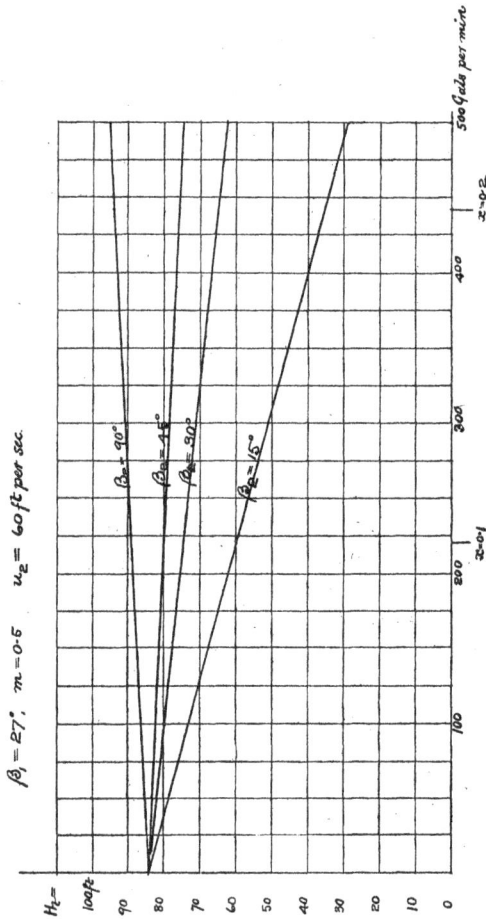

FIG. 10.—Curves showing variation of H'_s with Q: inlet angle β_1 constant.

The radial component of flow at inlet $= w_1 \sin \beta_1 = \dfrac{Q_t}{F_1}$

"   "   "   "   " outlet $= w_2 \sin \beta_1 = \dfrac{Q_t}{F}$

If $\qquad m = \dfrac{r_1}{r_2}$, $q = \dfrac{w_1}{w_2}$, and if $x = \dfrac{w_2 \sin \beta_2}{u_2}$,

and with the aid of the velocity diagrams at inlet and outlet, equation (1) may be written as: where $a = 1 - m^2$

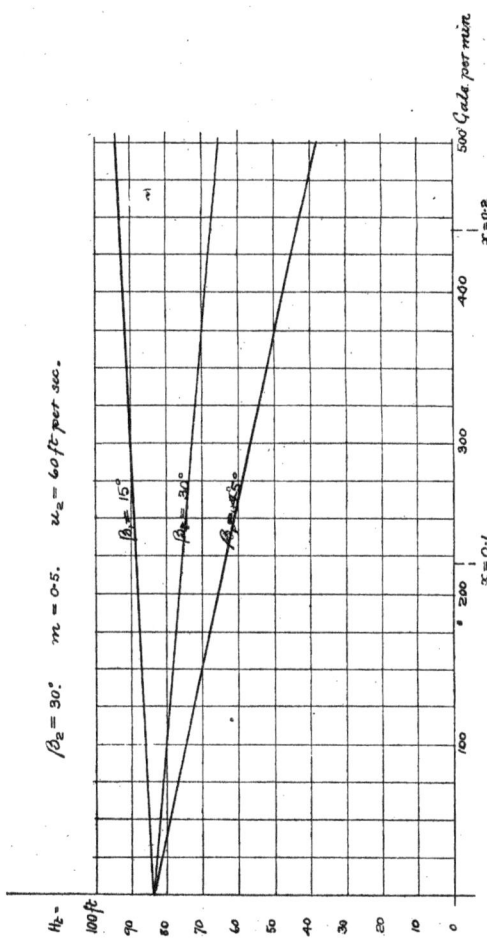

FIG. 11.—Curves showing variation H_t with Q_t when $\beta_2 = 90°$ and inlet angles β_1 are 15°, 30°, and 45°.

$$H_t = \frac{u_2^2}{g}\left\{ a - \left(\frac{\cos \beta_2}{\sin \beta_2} - mq\, \frac{\cos \beta_1}{\sin \beta_2}\right)x \right\}$$

and since x is a function of Q_t, this equation will show for a

2

given runner how H_t varies with Q_t, also how H_t will vary with u_2 when Q_t (or x) is constant.

Fig. 10 shows the variation of H_t with x (and Q_t) for runners having the values of $m = 0.5$, $\beta_1 = 27°$, $q = 1$,

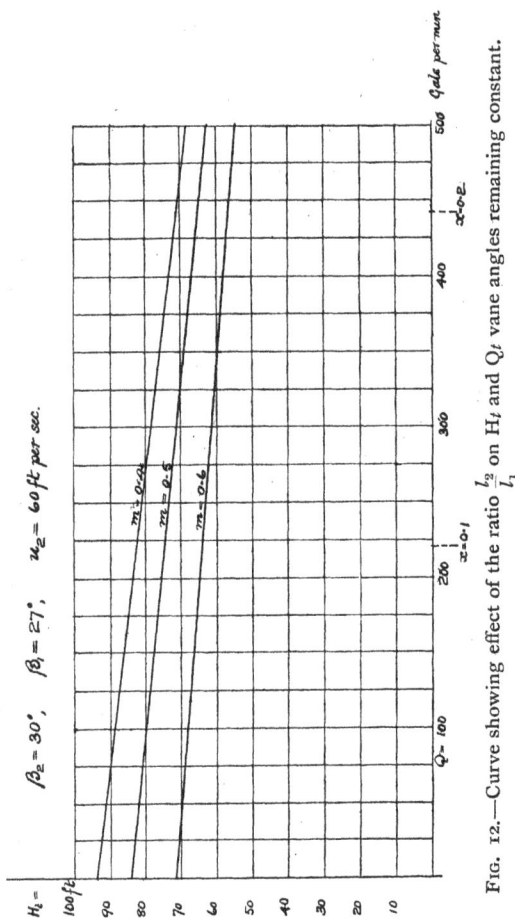

FIG. 12.—Curve showing effect of the ratio $\frac{b_2}{l_1}$ on H_t and Q_t vane angles remaining constant.

$b_2 = 0.1r_2$, $u = 60$ ft. per sec., for values of $\beta_2 = 15°$, $30°$, $45°$, and $90°$.

Fig. 11 shows the effect of the variation of H_t for runners having an angle of $\beta_2 = 30°$, but with values of $\beta_1 = 45°$, $30°$, $15°$, m and a being the same as above, but q being given by

$$q = \frac{1}{m} \frac{l_2}{l_1} \frac{\sin \beta_2}{\sin \beta_1}, \text{ where } \frac{l_2}{l_1} = \frac{1}{1 \cdot 75}.$$

$$\text{For } \beta_1 = 45°, \ q = 0 \cdot 81,$$
$$\beta_1 = 30°, \ q = 1 \cdot 14,$$
$$\beta_1 = 15°, \ q = 2 \cdot 2.$$

In Fig. 12, β_2 and β_1 are constant and the ratio $\frac{r_1}{r_2}$ or m varied. From these curves it will be seen how the pressure at zero delivery increases as the ratio $\frac{r_1}{r_2}$ gets smaller.

CHAPTER III.

HYDRAULIC LOSSES DURING THE PASSAGE OF FLUID THROUGH PUMP.

IN centrifugal or turbine pumps the theoretical head H_t is not all available for lifting the fluid, because of losses which occur during the fluid's passage through the pump.

These losses which affect the head are hydraulic losses and are enumerated as follows :—

1. The loss by shock due to the thickness of vanes at entrance to runner.

If just before entry to runner the component of v_1', in the direction of the vanes is w_1', immediately after entry w_1' is increased to w_1, due to the diminution of area caused by the vanes.

This sudden increase in velocity causes a loss due to shock, and if this loss be denoted by l_1

Then
$$l_1 = \frac{(w_1 - w_1')^2}{2g}.$$

If a_1' = area of passage at right angles to the direction of of w_1', neglecting vanes,

a_1 = area of passage at right angles to w_1 allowing for vanes. See Fig. 13.

b_1 = width of passage at right angles to plane of paper.

$a_1' = x b_1 \sin \beta_1$

$a_1 = p b_1 \sin \beta_1 = b_1 (x \sin \beta_1 - t \sin \beta_1)$

$\quad = b_1 \sin \beta_1 (x - t)$

$w_1 = w_1' \dfrac{a_1'}{a_1}.$

$\therefore l_1 = \dfrac{w_1'^2}{2g} \left(\dfrac{a}{a_1} - 1 \right)^2.$

(20)

This loss l_1 can be reduced by sharpening the vanes at inlet, but cannot be entirely eliminated, because the stream contracts in the first element of the vanes, therefore w_1 and a_1 should be taken as the velocity and cross section of the contracted stream.

2. The second loss is the loss of head due to friction whilst the fluid traverses the runner passages. This loss reduces the pressure energy that is gained if flow takes place through a passage of increasing cross section.

M. Hanocq shows that this loss is

$$l_2 = (1 - \phi^2)\left(2 + \frac{1}{q}\right)\frac{w_1^2}{2g} + (1 - \phi^2)\frac{w_2^2 - w_1^2}{2g}.$$

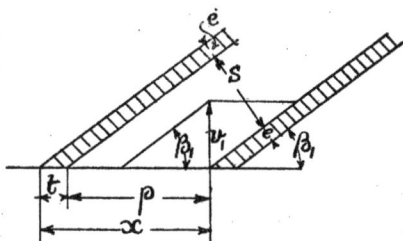

Fig. 13.—Diagram of vane element at entrance.

If the loss l_1 be included as being equal to $k\frac{w_1^2}{2g}$,

Then

$$l_1 + l_2 = (1 - \phi^2)\left(2 + \frac{1}{q}\right)\frac{w_1^2}{2g} + (1 - \phi^2)\frac{w_2^2 - w_1^2}{2g} + k\frac{w_1^2}{2g}$$

$$= \left[(1 - \phi^2)\left(2 + \frac{1}{q}\right) + k\right]\frac{w_1^2}{2g} + (1 - \phi^2)\frac{w_2^2 - w_1}{2g}.$$

Where ϕ is a constant depending on the dimensions of the passages, the angles β_1 and β_2, and the coefficient of friction and number of vanes, etc.,

$$q = \frac{w_1}{w_2}.$$

If $[(1 - \phi^2)\left(2 + \dfrac{1}{q}\right) + k]$, which is a constant, also depending on the dimensions of the passages, be written as equal to $1 - \psi^2$,

Then

$$l_1 + l_2 = (1 - \psi^2)\frac{w_1^2}{2g} + (1 - \phi^2)\frac{w_2^2 - w_1^2}{2g}.$$

This expression shows that the losses due to friction of the fluid flowing through the runner passages, and the eddy loss l_1 at entry to the passages, are functions of two coefficients which are connected together, the one affecting the kinetic energy at inlet and the other the pressure energy acquired by the fluid while flowing through the passages.

For runners of 200 mm. diameter, and with a width $0 \cdot 1 r_2$ or 10 mm. the following values of ϕ and ψ for angles $\beta_2 = 12°$, 30° and 90° respectively * :—

$\beta_2 = 12°$	$\beta_2 = 30°$	$\beta_2 = 90°$
$\beta_1 = 25°$	$\beta_1 = 35°$	$\beta_1 = 45°$
$\phi = 0\cdot92$	$\phi = 0\cdot95$	$\phi = 0\cdot96$
$\psi = 0\cdot67$	$\psi = 0\cdot82$	$\psi = 0\cdot89$
$q = 0\cdot706$	$q = 1\cdot25$	$q = 2$

3. Loss due to vane thickness at runner outlet. At outlet from runner the absolute velocity v_2 is suddenly reduced to the value v_2', in the clearance space between the runner and guides. This loss takes place at the expense of the pressure head, and will be denoted by l_3.

Then
$$l_3 = \frac{(v_2 - v_2')^2}{2g}.$$

This loss depends on the number and thickness of the vanes, and can be reduced by sharpening the vanes on the back as in Fig. 14.

4. Loss of head at entry to guide vanes due to their thickness.

* Values given by M. Hanocq in his work " Les Pompes Centrifuges ".

The absolute velocity in the clearance space between the tips of the runner vanes and the tips of the guide vanes is v_2', and the velocity at entry to guide vanes is v_3, then the loss of head due to sudden change in velocity is

$$l_4 = \frac{(v_3 - v_2')^2}{2g}.$$

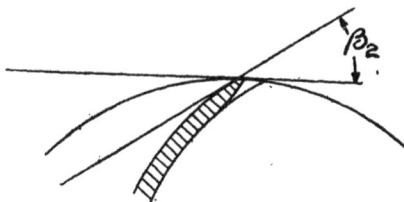

Fig. 14.—Sharpening of vane tip at runner outlet.

This loss is also reduced by sharpening the tips of the vanes as in Fig. 15.

5. This loss is the loss of head due to friction during the passages of the fluid through the guide or diffuser passages causing a reduction in the gain of pressure.

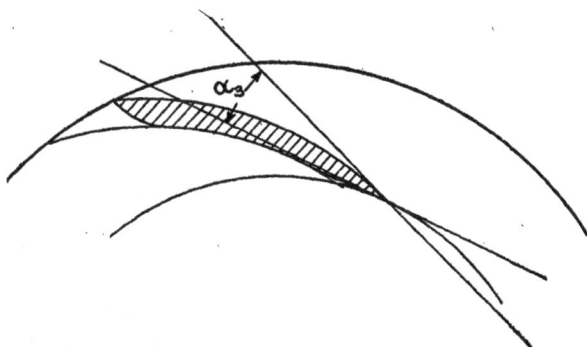

Fig. 15.—Entrance guide vane angle.

This loss is given by an expression similar to loss 2. M. Hanocq gives for this loss

$$l_5 = (1 - \psi^2)\frac{v_3^2}{2g} + (1 - \phi^2)\frac{v_4^2 - v_3^2}{2g}.$$

This may be expressed as a function of v_3, where v_3 is the velocity at inlet to diffuser.

If
$$\frac{v_3}{v_4} = n,$$

$$l_5 = (1 - \psi^2)\frac{v_3}{2g} + (1 - \phi^2)\frac{\left(\frac{v_3}{n}\right)^2 - v_3^2}{2g}$$

$$= (1 - \psi^2) + (1 - \phi^2)\left[1 - \left(\frac{1}{n}\right)^2\right]\frac{v_3^2}{2g} = (1 - \psi_1^2)\frac{v_3^2}{2g},$$

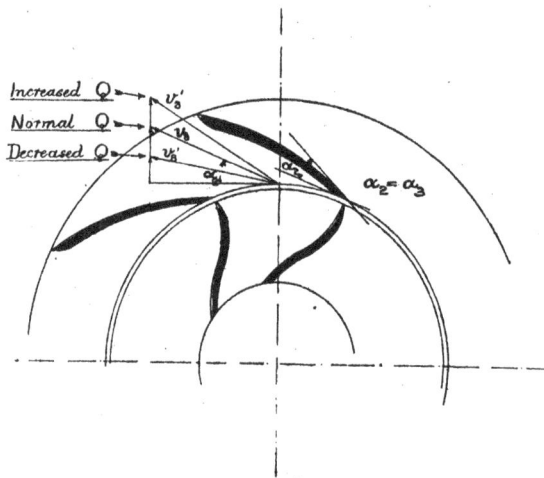

FIG. 16.—Effect of variation in quantity pumped.

the values of ψ, ϕ and ψ depending on the coefficient of friction and the dimensions of the passages.

6. In a pump having a diffuser with fixed guide vanes, and driven *at a constant speed*, there is a loss which occurs whenever the pump delivers a greater or a smaller quantity of fluid than the normal quantity.

This is due to the alteration in direction and magnitude of the absolute velocity v_3 at entrance to the guide vanes, with varying quantities, the angle a_3 being fixed and correct for one quantity only, i.e. normal quantity.

This alteration in direction causes a loss of head due to impact on one side of the guide vanes.

This action is shown in Fig. 16, when Q is in excess of normal, absolute velocity is v_3' and there is impact on the convex side of the guide vanes.

When the quantity is less than normal, there is impact on the concave side of the guide vanes.

Also a loss due to sudden change of velocity v_3' at entry to the guide vanes equal to $\dfrac{(v_3 - v_3')^2}{2g}$.

The loss due to impact, see Fig. 17, is $\dfrac{(v_3' \sin \theta)^2}{2g}$.

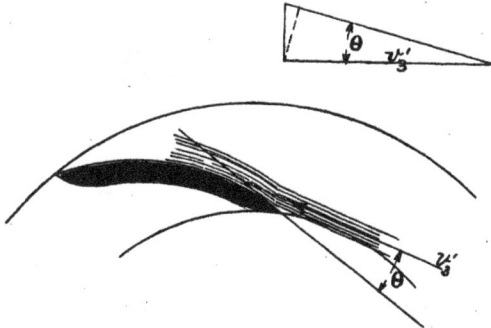

FIG. 17.—Impact of stream on guide vane.

Whether the above is always true is doubtful, especially in the case of quantities below normal.

In this case there is probably very little change in the velocity, the fluid flowing as indicated in Fig. 18, the impact loss being the only one, whereas the above would be nearer the truth for quantities greater than normal.

The test curves connecting Q and H for pumps with fixed guide vanes seem to confirm the latter.

7. The loss of head due to friction in the casing will be denoted by l_7 and may be taken as

$$l_7 = \zeta \frac{v_c^2}{2g}$$

ζ being a co-efficient, and v_c the velocity of flow in the casing.

With the spiral casing of gradually increasing cross section v_c is constant, and as an approximation, the casing will be assumed to be a taper bend of average diameter d, and average radius of curvature R of mean path A$bcef$ (Fig. 36).
l_7 may then be written as

$$t_7 = \zeta_b \frac{4l}{d} \frac{v_c^2}{2g}.$$

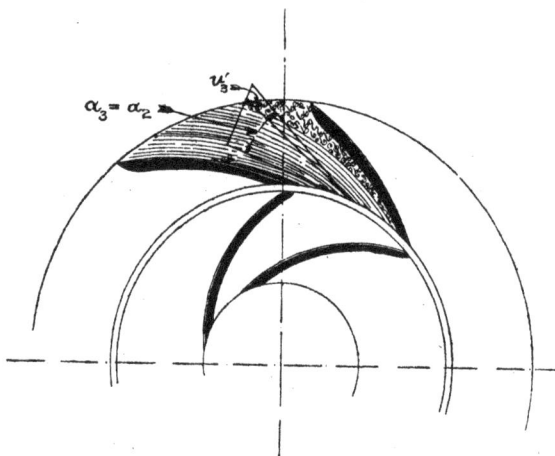

FIG. 18.—Effect of eddies in guide channels.

ζb = co-efficient which depends on the curvature of the path A$bcef$ (Fig. 36).

l = length of mean path in feet A$bcef$ (Fig. 36).

d = diameter of pipe in feet, which will be taken as average diameter of pipe (Fig. 36).

Prof. Unwin gives the following as some values of ζ_b for various values of $\frac{R}{d}$:—

$\frac{R}{d}$	20	10	5	3	2
ζ_b	0·004	0·008	0·016	0·032	0·048

In the case of a casing concentric with the runner, v_c is not constant owing to different quantities passing the cross sections of constant area.

In this case the estimation of l_7 is difficult, but this loss is kept small by making v_c as small as possible.

8. This loss is the loss by shock at entry to the casing, and if v is the velocity of fluid at entry to casing, and v_c is the velocity in the casing,

$$l_8 = \frac{(v - v_c)^2}{2g}.$$

CHAPTER IV.

MANOMETRIC HEAD, CHANGE OF PRESSURE, AND HYDRAULIC EFFICIENCY OF CENTRIFUGAL PUMPS.

FIG. 19 shows roughly the general arrangement of a centri-fugal pump.   H is the total height to which the water is to be

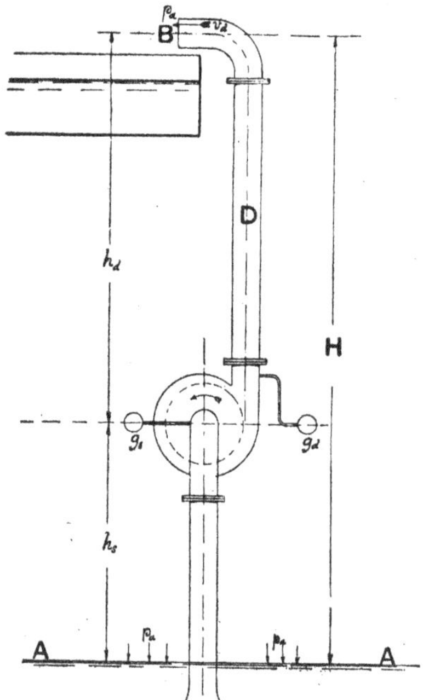

FIG. 19.—Typical centrifugal pump installation.

lifted, that is the potential energy to be given to each lb. of fluid.

h_s is the height of centre of pump above the level of the fluid in the suction tank.   Theoretically for water, h may be

(28)

anything up to 32 ft., but due to air leakage and friction in suction pipe is much less than this.

h_d is the height of the centre of delivery outlet above the centre of the pump, Fig. 19.

Then $H = h_s + h_d$ or what is termed the static head.

The velocity of flow in delivery pipe is v_d, therefore the kinetic energy of the fluid is $\dfrac{v_d^2}{2g}$, and this is wholly lost.

If l_s and l_d represent the loss of head due to friction in suction and delivery pipes, and if p_1 and p_c be the reading of the two pressure gauges g_s and g_d on the pump suction at inlet to runner, and at outlet from diffuser or pump casing, the pressure at the level of the fluid at AA is p_a, the pressure at B is also p_a, which is, of course, that of the atmosphere, and the hydraulic efficiency of the whole installation is $\dfrac{H}{H_t}$ and the total over-all efficiency is $\dfrac{QHG}{550N_p}$.

(a) *Pump without Guide Vanes, Whirlpool Chamber, or Volute, or one having a Badly Designed Volute or Casing.*

In a pump of this type, illustrated in Fig. 5, the greater part of the kinetic energy $\dfrac{v_2^2}{2g}$ per unit weight of fluid at outlet from the runner is dissipated by the eddies caused by shock.

Considering the suction pipe from the level AA, up to the inlet of the runner, Fig. 19, the velocity at AA will be very small, and may be taken as equal to 0, the velocity at entry to runner being v_1.

Then on applying Bernoulli's theorem to the flow in the suction pipe up to the inlet or eye of the pump runner

$$\frac{p_a}{G} = \frac{p_1}{G} + h_s + \frac{v_1^2}{2g} + l_s$$

or

$$\frac{p_1}{G} - \frac{p_a}{G} + h_s + \frac{v_1^2}{2g} + l_s = 0 \qquad (a)$$

Applying the same law to the flow from the pump casing up to the section at B, Fig. 19, or in other words the flow in the delivery pipe,

$$\frac{p_c}{G} + \frac{v_c^2}{2g} = \frac{p_a}{G} + h + \frac{v_d^2}{2g} + l$$

or

$$\frac{p_a - p_c}{G} + h_d + \frac{v_d^2}{2g} - \frac{v_c^2}{2g} + l_d = 0 \qquad . \qquad . \quad (b)$$

v_c = velocity in the casing and p_c the corresponding pressure. Adding (a) and (b),

$$\frac{p_1 - p_c}{G} + h_d + h_s + l_d + l_s + \frac{v_d^2}{2g} - \frac{v_c^2}{2g} + \frac{v_1^2}{2g} = 0$$

$$(h_d + h_s) + (l_d + l_s) = \frac{p_c - p_1}{G} + \frac{v_c^2}{2g} - \frac{v_d^2}{2g} - \frac{v_1^2}{2g}.$$

The quantity $h_d + h_s = $ H, or what is termed the static head; also in this type of pump the pressure p_c is equal to p_2, the pressure at runner outlet, there being no increase of pressure due to flow into casing, therefore,

$$H + l_s + l_d = \frac{p_2 - p_1}{G} + \frac{v_c^2}{2g} - \frac{v_d^2}{2g} - \frac{v_1^2}{2g} \qquad . \qquad . \quad (c)$$

Also the total term on the left-hand side of the equation, namely $(h_s + h_d) + (l_s + l_d)$, or $H + l_s + l_d$, is the head given by the readings of two manometers, the one on the suction near the eye of the pump and the other one fixed at the outlet from the diffuser or volute.

This quantity $H + l_d + l_s$ is called the manometric head H_m.

Then equation (c) will become

$$H_m = \frac{p_2 - p_1}{G} + \frac{v_c^2}{2g} - \frac{v_d^2}{2g} - \frac{v_1^2}{2g} . \qquad . \qquad . \quad (d)$$

The hydraulic efficiency of the pump only will be denoted η_h.

or

$$\eta_h = \frac{H_m}{H_t}$$

and $H_m = \eta^h H_t$.

The change of pressure of the fluid in a pump of this type is the change of pressure in the runner only and is the term

$$\frac{p_2 - p_1}{G}.$$

During the passage of the fluid through the runner the change of pressure energy is equal to the change of pressure due to centrifugal force of a particle between inlet and outlet, together with the change in pressure which occurs when a fluid flows in a passage of varying cross section.

Therefore,

$$\frac{p_2 - p_1}{G} = \frac{u_2^2 - u_1^2}{2g} + \frac{w_1^2 - w_2^2}{2g} - l_1 - l_2 - l_3,$$

p_1 and p_2 being the pressures per sq. foot at runner inlet and outlet;

u_1 and u_2 being the peripheral velocities at runner inlet and outlet;

w_1 and w_2 being the relative velocities of the fluid at inlet and outlet of the runner, and l_1, l_2, l_3 being losses of head due vane thickness and friction.

Substituting this value for $\dfrac{p_2 - p_1}{G}$ in terms of the velocities in equation (d),

$$H_m = \frac{u_2^2 - u_1^2}{2g} + \frac{w_1^2 - w_2^2}{2g} + \frac{v_c^2}{2g} - \frac{v_d^2}{2g} - \frac{v_1^2}{2g} - l_1 - l_2 - l_3 \quad (e)$$

Again, in this type of pump the velocity in the casing is usually the same as the velocity in the delivery pipe, in this case $v_c = v_d$, and equation (e) becomes

$$H_m = \frac{u_2^2 - u_1^2}{2g} + \frac{w_1^2 - w_2^2}{2g} - \frac{v_1^2}{2g} - l_1 - l_2 - l_3.$$

(b) Pump with Spiral Chamber or Volute.

If the pump has a well-designed volute or spiral chamber, there is a gain of hydraulic efficiency compared with the previous type, owing to part of the kinetic energy at discharge from runner being converted into pressure energy.

Fig. 20 shows a casing of this type of pump (Fig. 4).

In this chamber the cross section gradually increases from A to D in such a manner that the velocity of flow v_c is constant.

The angle between the tangent to the mean line of flow in

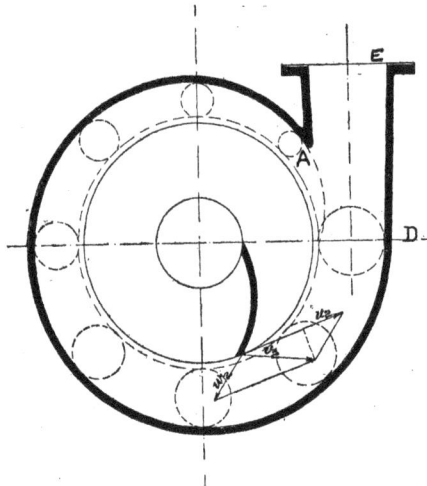

FIG. 20.—Spiral chamber or volute.

the volute and v_2 is practically equal to a_2, and will be taken as such.

At outlet from runner the absolute velocity v_2 is suddenly changed to that of the volute v_c; consequently, there is a loss of head due to impact.

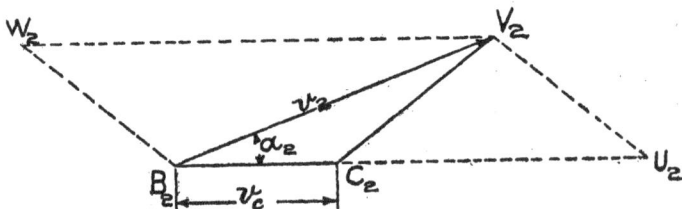

FIG. 21.—Diagram for loss at entry to spiral chamber.

The change in velocity is represented by the length C_2V_2 in Fig. 21, and the loss of head is

$$\frac{(C_2V_2)^2}{2g} = \frac{v_2^2 + v_c^2 - 2v_2v_c \cos a_2}{2g}.$$

Again, if the velocity v_2 were gradually changed to that of v_c, the gain in pressure energy would be $\dfrac{v_2{}^2 - v_c{}^2}{2g}$, but as the velocity v_2 is suddenly changed to that of v_c the gain in pressure head, if p_2 = pressure at runner outlet, and p_c = pressure in volute, is therefore

$$\frac{p_c - p_2}{G} = \frac{v_2{}^2 - v_c{}^2}{2g} - \frac{(C_2 V_2)^2}{2g}$$

$$= \frac{v_2{}^2 - v_c{}^2}{2g} - \frac{v_2{}^2 + v_c{}^2 - 2v_2 v_c \cos a_2}{2g}$$

$$= \frac{2(v_2 v_c \cos a_2 - v_c{}^2)}{2g}.$$

The gain in pressure head is a maximum when

$$v_c = \frac{v_2 \cos a_2}{2}.$$

In which case

$$\frac{p_c - p_2}{G} = \frac{v_2{}^2 \cos^2 a_2}{4g}.$$

A still further gain of pressure and increase in hydraulic efficiency may be obtained by having a taper discharge pipe of increasing diameter, which gradually reduces the velocity v_c in the volute, to a smaller value v_d in the discharge pipe.

The gain in pressure energy from D to E, Fig. 20, if p_d represents the pressure at E in discharge pipe is

$$\frac{p_d - p_c}{G} = \frac{v_c{}^2 - v_d{}^2}{2g}.$$

The total change of pressure energy in a pump of this type with a taper discharge pipe is

$$\frac{p_d - p_1}{G} = \frac{u_2{}^2 - u_1{}^2}{2g} + \frac{w_1{}^2 - w_2{}^2}{2g} + \frac{2(v_2 v_c \cos a_2 - v_c{}^2)}{2g} - \frac{v_d{}^2}{2g}$$
$$- l_1 - l_2 - l_3 - l_7.$$

If the casing is not fitted with a taper discharge pipe then

$$p_d = p_c.$$

3

The change of pressure energy in this case is

$$\frac{p_c - p_1}{G} = \frac{u_2^2 - u_1^2}{2g} + \frac{w_1^2 - w_2^2}{2g} + \frac{2(v_2 v_c \cos a_2 - v_c^2)}{2g} - l_1 - l_2 - l_3 - l_7.$$

Applying Bernoulli's theorem as in the previous case

$$\begin{aligned}
H_m &= \frac{p_d - p_1}{2g} - \frac{v_1^2}{2g} \\
&= \frac{u_2^2 - u_1^2}{2g} + \frac{w_1^2 - w_2^2}{2g} + \frac{2 v_2 v_c \cos a_2}{2g} - \frac{v_c^2}{2g} - \frac{v_d^2}{2g} - \frac{v_1^2}{2g} \\
&\qquad\qquad\qquad\qquad\qquad\qquad - l_1 - l_2 - l_3 - l_7,
\end{aligned}$$

for pump with a taper discharge pipe.

If the pump is not fitted with a taper discharge $p_d = p_c$ and the velocity in the casing v_c is the same as the velocity in the discharge pipe.
Then

$$H_m = \frac{p_c - p_1}{G} - \frac{v_1^2}{2g}.$$

And

$$\frac{p_c - p_1}{G} = \frac{u_2^2 - u_1^2}{2g} + \frac{w_1^2 - w_2^2}{2g} + \frac{2(v_2 v_c \cos a_2 - v_c^2)}{2g} - l_1 - l_2 - l_3 - l_7.$$

Therefore

$$H_m = \frac{u_2^2 - u_1^2}{2g} + \frac{w_1^2 - w_2^2}{2g} + \frac{2(v_2 v_c \cos a_2 - v_c^2)}{2g} - \frac{v_1^2}{2g}$$
$$- l_1 - l_2 - l_3 - l_7$$

$$\eta_h = \frac{H_m}{H_t} \qquad \text{or } H_m = \eta_h H_t$$

l_1, l_2, l_3, and l_7 being loss of head as described in Chapter III.

(c) Pump with Whirlpool Chamber.

In Fig. 22 the runner is surrounded by a concentric chamber having parallel sides, called a whirlpool chamber, through which the fluid flows before passing into the volute.

The fluid at exit from the runner is discharged freely into this chamber, forming a free spiral vortex, the fluid describing stream lines having a constant inclination to the tangent of concentric surfaces of the whirlpool chamber.

Fig. 23 shows the path AB taken by the elementary streams flowing through the whirlpool chamber.

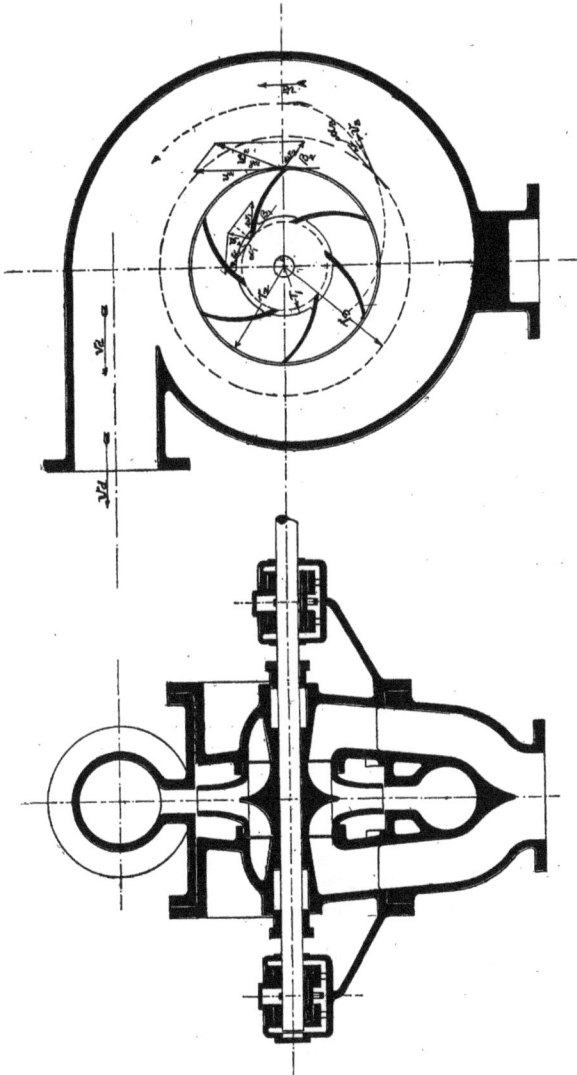

FIG. 22.—Centrifugal pump with whirlpool chamber.

In such a vortex the energy along the stream is constant

(neglecting friction which tends to decrease the pressure head),

or $\dfrac{P}{G} + \dfrac{v^2}{2g} = \text{constant, and } v \propto \dfrac{1}{r}.$

If r_2 and r_3 are the inner and outer radii of the chamber in feet,

p_2 and p_3 are the pressures at r_2 and r_3,

v_2 and v_3 „ „ velocities „ „ „

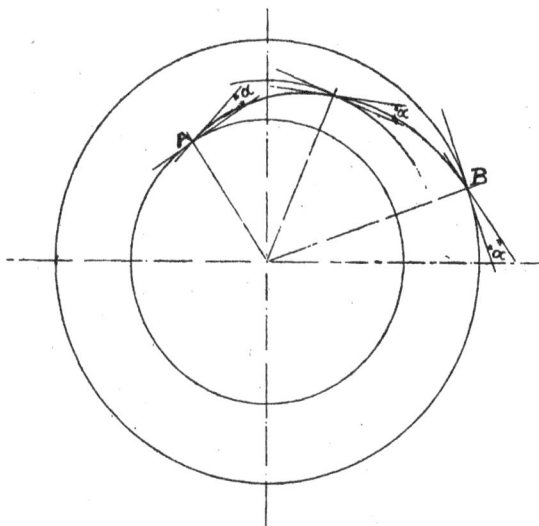

FIG. 23.—Path described by water particles in whirlpool chamber.

The increase of pressure energy is

$$\frac{p_3 - p_2}{G} = \frac{v_2^2}{2g} - \frac{v_3^2}{2g},$$

$$v_3^2 = \left(\frac{v_2 r_2}{r_3}\right)^2,$$

$$\frac{p_3 - p_2}{G} = \frac{v_2^2}{2g} - \frac{v_2^2}{2g}\frac{r_2^2}{r_3^2} - \text{friction losses,}$$

$$= \frac{v_2^2}{2g}\left(1 - \frac{r_2^2}{r_3^2}\right) - \text{friction losses.}$$

The gain in pressure head increases as $\dfrac{r_2^2}{r_3^2}$ diminishes or as r_3 increases, which means as the diameter of the casing increases, and consequently the weight.

In the case of pumps whose capacity is small compared with the head pumped against, the absolute velocity of the fluid at exit from the runner is high and consequently the angle a_2 is very small.

When a_2 is very small, the length of the path of the stream lines in a free spiral vortex is very long, and since the loss due to friction is proportional to the length of the path and the square of the velocity, the efficiency of a whirlpool chamber as a converter of kinetic into pressure energy is considerably reduced.

In pumps of small capacity it is therefore essential that, instead of a whirlpool chamber or plain diffuser with parallel sides, a diffuser fitted with guide vanes should be employed, because the width of the runner at exit should not be less than $0.1 r_2$, on account of friction losses, and this makes the angle a_2 sometimes as small as $5°$.

With an angle a_2 as small as $5°$, the path of the water particles, for a given diminution of velocity, will be very long in a plain diffuser or whirlpool chamber, and consequently the friction losses will be high for such a pump.

In order to reduce these losses, guide vanes are used, whose object is to shorten the path of the stream lines. To fulfil this condition efficiently, the cross sections of the passages formed by two consecutive vanes and the diffuser walls should not diverge too rapidly, or, due to eddy losses, the gain of pressure energy expected will not be obtained.

Fig. 22 shows in diagrammatic form a pump with a whirlpool chamber.

The total change of pressure energy in a pump of this type is equal to the change of pressure in the runner + change of pressure in diffuser or whirlpool chamber + change of pressure from whirlpool chamber to volute + change of pressure in discharge pipe.

$$\frac{p_d - p_1}{G} = \frac{u_2^2 - u_1^2}{2g} + \frac{w_1^2 - w_2^2}{2g} + \frac{v_2^2}{2g}\left(1 - \frac{r_2^2}{r_3^2}\right) + \frac{2(v_3 v_c \cos a_3 - v_c^2)}{2g}$$
$$+ \frac{v_c^2 - v_d^2}{2g} - l_1 - l_2 - l_3 - l_7 - l_w.$$

Again, applying Bernoulli's theorem as before,

$$H_m = \frac{u_2^2 - u_1^2}{2g} + \frac{w_1^2 - w_2^2}{2g} + \frac{v_2^2}{2g}\left(1 - \frac{r_2^2}{r_3^2}\right) + \frac{2(v_3 v_c \cos a_3 - v_c^2)}{2g}$$
$$+ \frac{v_c^2 - v_d^2}{2g} - \frac{v_1^2}{2g} - l_1 - l_2 - l_3 - l_7 - l_w,$$

for pump with a taper discharge pipe.

If no taper discharge pipe at exit to casing, then

$$H_m = \frac{u_2^2 - u_1^2}{2g} + \frac{w_1^2 - w_2^2}{2g} + \frac{v_2^2}{2g}\left(1 - \frac{r_2^2}{r_3^2}\right) + \frac{2(v_3 v_c \cos a_3 - v_c^2)}{2g}$$
$$- \frac{v_1^2}{2g} - l_1 - l_2 - l_3 - l_7 - l_w,$$

the hydraulic efficiency $\eta_h = \dfrac{H_m}{H_t}$ or $H_m = \eta_h H_t$,

l_1, l_2, l_3, etc., being losses of head as described in Chapter III.

The hydraulic efficiency of a pump of this type depends on the value of the ratio $\left(\dfrac{r_2}{r_3}\right)^2$; the smaller this ratio the greater the weight of the pump, and consequently its cost.

Pump with Guide Vanes in the Diffuser.

From the preceding paragraph it will be seen that when the quantity is small compared with the head, it is absolutely necessary to use guide vanes in the diffuser in order to obtain a high efficiency.

These should be so shaped that the velocity in the passages formed by two consecutive guide vanes decreases uniformly along their length, and that the calculated area of the cross sections of the passages is strictly adhered to in their construction. Fig. 2 shows a single stage pump of this type and Fig. 3 the runner and guide vanes of either a single stage or a multi-stage pump.

In order to prevent the tips of the guide vanes being damaged by any foreign matter, it is advisable to have a certain amount of clearance between the outer diameter of the runner and the inner diameter of the guide ring.

This clearance depends on the size of pump, but principally on the nature of the foreign matter in the fluid to be pumped.

In this clearance space shown in Fig. 3 and bounded by the radii r_2 and r_3, the path of the fluid particles will be equiangular spirals; therefore the angle a_3 of the tip of the guide vane will be equal to the angle a_2, which the absolute velocity of the fluid particles at exit from runner makes with the tangent to the runner periphery.

If p_3 and p_4 be the pressures at inlet and outlet of guide vanes, and v_3 and v_4 the velocities at the same points, then the increase of pressure energy in flowing through the guides is

$$\frac{p_4 - p_3}{G} = \frac{v_3{}^2 - v_4{}^2}{2g} - l_4 - l_5.$$

And the gain of pressure in the clearance space is

$$\frac{p_3 - p_2}{G} = \frac{v_2{}^2}{2g}\left(1 - \frac{r_2{}^2}{r_3{}^2}\right) - \text{loss}.$$

Since this clearance space is usually small this gain of head would be neutralised by the losses and therefore be neglected.

Applying Bernoulli's theorem as in previous cases, the manometric head is

$$H_m = \frac{p_d - p_1}{G} - \frac{v_1{}^2}{2g}.$$

For a pump of the type in Fig. 2 with volute and taper discharge pipe

$$\frac{p_d - p_1}{G} = \frac{u_2{}^2 - u_1{}^2}{2g} + \frac{w_1{}^2 - w_2{}^2}{2g} + \frac{v_3{}^2 - v_4{}^2}{2g} + \frac{2(v_4 v_c \cos a_4 - v_c{}^2)}{2g}$$

$$+ \frac{v_c{}^2 - v_d{}^2}{2g} - l_1 - l_2 - l_3 - l_4 - l_5 - l_6 - l_7,$$

$$H_m = \frac{u_2^2 - u_1^2}{2g} + \frac{w_1^2 - w_2^2}{2g} + \frac{v_3^2 - v_4^2}{2g} + \frac{2(v_4 v_c \cos a_4 - v_c^2)}{2g} + \frac{v_c^2}{2g}$$

$$- \frac{v_d^2}{2g} - \frac{v_1^2}{2g} - l_1 - l_2 - l_3 - l_4 - l_5 - l_6 - l_7.$$

If the pump has volute but no taper discharge pipe then $v_d = v_c$.

And
$$\frac{v_c^2}{2g} - \frac{v_d^2}{2g} = 0.$$

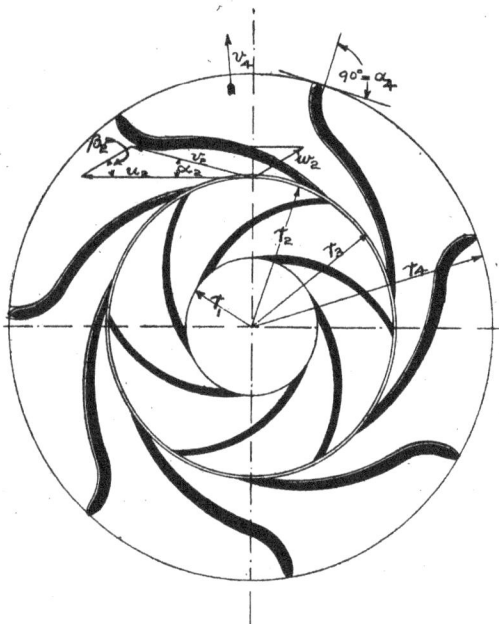

FIG. 24.—Runner with guide vanes for discharge into concentric chamber.

In multi-stage pumps there is no spiral chamber (unless in the last stage), and, further, the guide vanes are designed so that they discharge radially into a concentric chamber (see Fig. 24).

In this case, there will be no increase of pressure energy, such as takes place when the discharge is into a volute, in which the fluid is moving nearly in the same direction.

Therefore the increase of pressure energy represented by the term $2(v_4 v_c \cos a_4 - v_c^2)/2g$, will not be obtained and the manometric head in this case will be

$$H_m = \frac{v_2^2 - v_1^2}{2g} + \frac{v_{r_1}^2 - v_{r_2}^2}{2g} + \frac{v_0^2}{2g} \cdot \frac{v_4^2}{} - \frac{v_1^2}{2g}$$

$$- l_1 - l_2 - l_3 - l_4 - l_5 - l_6 - l_7.$$

As before $\qquad \eta_h = \dfrac{H_m}{H_t}$ or $H_m = \eta_h H_t$.

Also in this latter case the term l_7 would be the friction in the passages connecting the outlet of one stage with the inlet of the next stage.

Multi-stage Pumps.

If the head against which a centrifugal pump is to work is such that the speed is excessive, or that the quantity is so small compared with the head that the width of the passages is too small to obtain a reasonable efficiency, then two or more pumps must be used working in series, or at different levels.

In Fig. 25, scheme A represents two pumps working at different levels, being used to lift water from a low level C to a higher level E, total static head being equal to H; the first pump drawing water from tank C, and delivering it into a tank D, lifting the water through height H_1.

The second pump is placed at a higher level, drawing its water from tank D, into which the first pump delivers, and delivering into a tank E which is H_2 ft. or metres above tank D, and H ft. or metres above tank C. Both pumps are identical, delivering the same quantity, and running at the same speed.

For the first pump,

$$H_1 + l_{s_1} + l_{d_1} = \frac{p_{d_1} - p_{s_1}}{G} - \frac{v_1^2}{2g} = H_{m_1}.$$

For the second pump,

$$H_2 + l_{s_2} + l_{d_2} = \frac{p_{d_2} - p_{s_1}}{G} - \frac{v_1^2}{2g} = H_{m_2}.$$

But since the two pumps are the same in every respect,

$$\frac{p_{d_1} - p_{s_1}}{G} - \frac{1}{2g} = \frac{p_{d_2} - p_{s_2}}{G} - \frac{v_1^2}{2g}.$$

$$\therefore H_{m_1} = H_{m_2}.$$

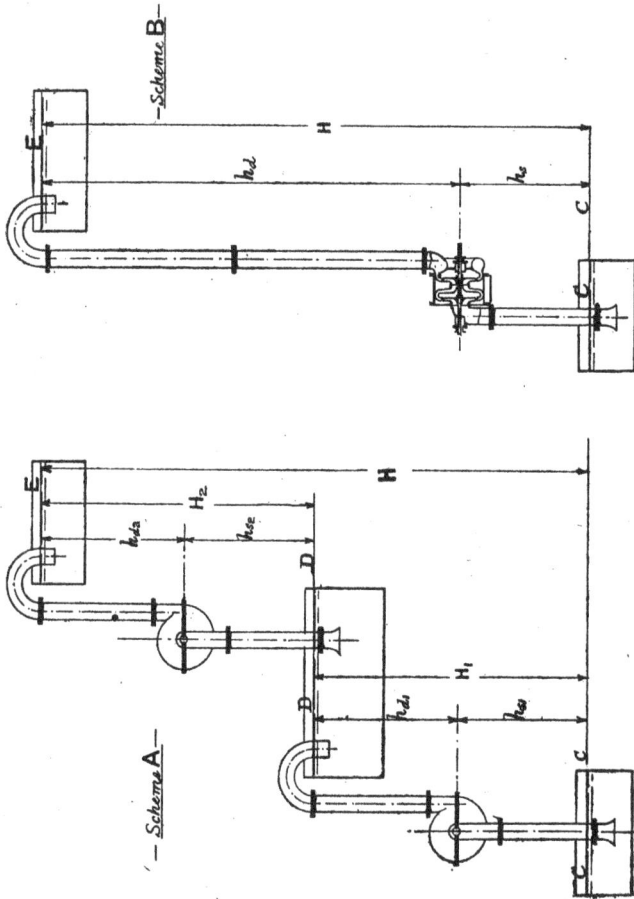

Multistage pump (2 stages).

FIG. 25.

Two single stage pumps in series,

Also it is easily seen from the above that,

$$H_1 + H_2 + l_{s_1} + l_{d_1} + l_{s_2} + l_{d_2} = 2\left(\frac{p_{d_1} - p_{s_1}}{G} - \frac{v_1^2}{2g}\right) = 2H_{m_1}$$

$$H_1 + H_2 + l_{s_1} + l_{d_1} + l_{s_2} + l_{d_2} = H + l_{s_1} + l_{d_1} + l_{s_2} + l_{d_2}$$

$$= H_m$$

From this it is obvious that if, instead of two separate pumps working at different levels as in scheme A, Fig. 25, two identical runners are keyed to the same shaft, each with its guide ring, the discharge from the first being conducted to the entry of the second runner, as in Scheme B, Fig. 25, the same total head will be overcome by this pump situated at a low level.

The advantage of the scheme B is that the cost would be less and the efficiency greater, owing to the reduction in number of the bearings, stuffing boxes, and glands, etc., also the advantage of only one unit in place of two.

Each runner with its respective diffuser (with or without vanes) constitutes what is known as a stage, a multi-stage pump being one with more than one stage.

Then for any multi-stage pump, the total manometric head Hm, if n be the number of stages, and Hm_1 the manometric head produced in one stage

$$Hm = nHm_1 \text{ or } Hm_1 = \frac{Hm}{n}.$$

The hydraulic efficiency is

$$\eta_h = \frac{Hm}{nHt_1} = \frac{Hm_1}{Ht_1}.$$

CHAPTER V.

Bearings.

THE bearings for carrying the shafts of centrifugal pumps do not present any great difficulties in their design.

The best form of bearing is a ring lubricated bearing lined with anti-friction metal, which should be independent and bolted on to the pump casing.

The essential points to bear in mind when designing these bearings are: to have the oil well of ample capacity, to have fairly heavy lubricating rings, in order to prevent chattering when running at high speeds; that the fluid pumped cannot enter bearing or oil well, and so wash out the oil; also care should be taken that whitemetal lining of the bush cannot work loose, this being prevented by casting dovetail grooves in the bush.

Fig. 26 represents one of the best forms of bearings for this class of machinery.

The white-metal lined cast-iron bush is in halves, the bottom half supporting the shaft with its whole length; in the top half are stop pins to prevent the bush from turning, also allowing the bottom half of bush to be removed without lifting the shaft.

The oil ring, in halves, and fastened by screws, is of trapezoidal cross section and is made of bronze.

A flyer is on the shaft which throws any water off, by centrifugal force, which leaks from the gland along the shaft, thus preventing it entering the bearing; also, in a similar manner, preventing the lubricant from working out of the bearing housing.

(44)

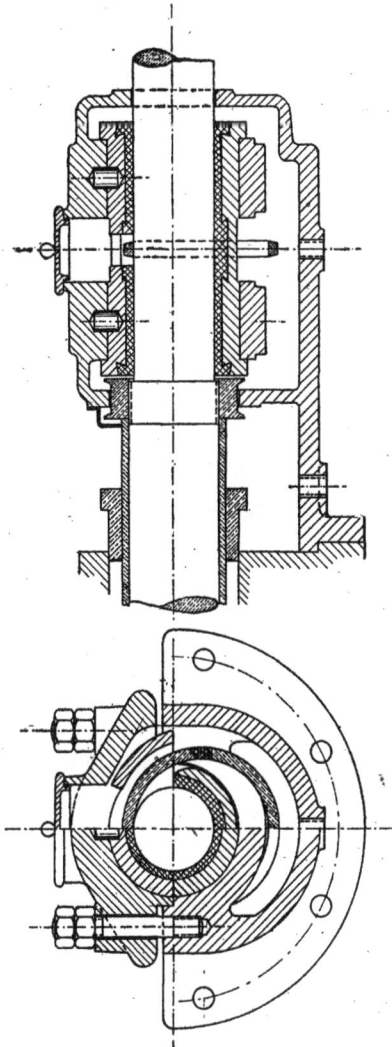

FIG. 26.—Bearing.

For speeds of from 3000 to 6000 revolutions per minute, it is advisable to use self-aligning or ball-seated bearing steps, as illustrated in Fig. 27.

The work expended in overcoming the friction of the shaft journal in the bearing is transformed into heat energy.

This heat must be dissipated by radiation at such a rate that the temperature of the bearing surface does not exceed a reasonable value.

The work expended in overcoming the friction of the shaft journal in its bearing, in ft. lb. per second is

$$\mu p l d c.$$

μ = coefficient of friction.

p = intensity of pressure on the projected area of the journal in lb. per square inch.

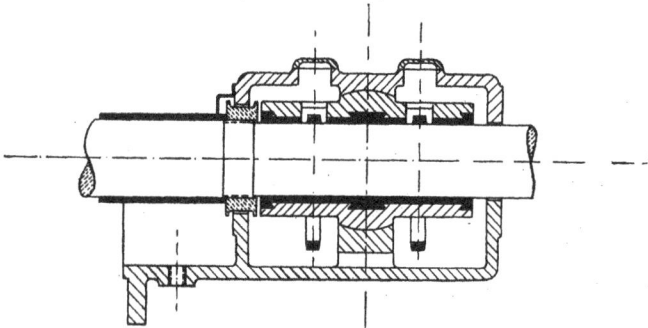

FIG. 27.—Self-aligning bearing.

d = diameter of the journal in inches.

l = length of journal in inches.

c = velocity of rubbing surface in feet per second.

The heat generated per hour by friction in B.T.Us. is

$$\frac{\mu p l d c}{778} \times 3600,$$

and must be radiated by the bearing.

Lasche found that the coefficient of friction μ depended on the temperature of the rubbing surfaces, provided that p was between the limits of from 15 to 200 lb. per sq. in., and t was between the limits of from 80° to 212° F.

The relation between these quantities, under the above conditions, is given by the equation

$$\mu p(t - 32°) = \text{constant} = 51\cdot2.$$

Lasche also found that the rate at which the heat is radiated per hour depends on the type of bearing and the air circulation.

If Q_b = heat radiated per hour by bearing,

t = temperature of the rubbing surfaces,

t_0 = temperature of the atmosphere,

$$Q_b = \frac{1}{778}k\mu dl(t - t_0).$$

The constant k depends on the type of bearing, and may be taken as equal to 50 for the types illustrated.

Q_b should be equal to, or greater, than the work of friction in heat units per hour.

If Q_b is found to be less than the friction work, then water cooling of the bearing, or forced lubrication, must be resorted to.

In the bearings illustrated, the general practice is to make the length of bearing l equal to three times its diameter d.

Assuming that $l = 3d$, and that t is equal to 120° F., then the horse-power necessary to overcome bearing friction may be estimated when designing a pump bearing.

If this horse-power is denoted by N_b and if n is the number of revolutions per minute,

$$N_b = 0\cdot00001385d^3n.$$

Power Absorbed and Over-all Efficiency.

The power expended in giving each pound of fluid passing through the pump runner energy equal to H_t ft. lb., will be called the theoretical power, and will be denoted by N_t.

Also, due to short-circuit losses, or leakage losses due to difference of pressure on both sides of the running joints, the quantity passing through the discharge pipe is less than the quantity passing through the runner.

If Q = quantity passing through the pump runner per second,

Q_a = quantity passing through the discharge pipe per second,

L = quantity short-circuited, quantity lost due to leakage $Q_t = Q_a + L$.

And

$$N_t = \frac{H_t Q_t G}{550} = \frac{H_t (Q_a + L) G}{550}.$$

In order to estimate the power necessary to drive the pump, the power required to overcome the friction of the two outside faces of the runner against the surrounding fluid, and the frictional resistance of the stuffing boxes and bearings must be taken into account, and added to the theoretical power N_t.

Although the power necessary to overcome disc friction does not affect the head pumped against, it has considerable influence on the power required to drive the pump, and if a high efficiency is to be obtained it must be kept as low as possible.

If N_r denotes the power required to overcome the friction of the two sides of the runner rotating in the surrounding fluid, then for all practical purposes,

$$N_r = C u_2^3 r_2^2.$$

Where u_2 = peripheral velocity of runner.

r_2 = outer radius of runner in feet.

c = constant, which for want of a better value will be taken as 0·0000087.

Therefore $\qquad N_r = 0·0000087 u_2^3 r_2^2.$

The chart, Fig. 29, enables N_r to found for different diameters of runners revolving at various speeds of rotation.

Assuming that the bearings are three diameters long and that the pump has two bearings of the same diameter, the temperature of the rubbing surfaces being 120° F. ;

The horse power to overcome journal friction is equal to 0·0000277 $d^3 n$;

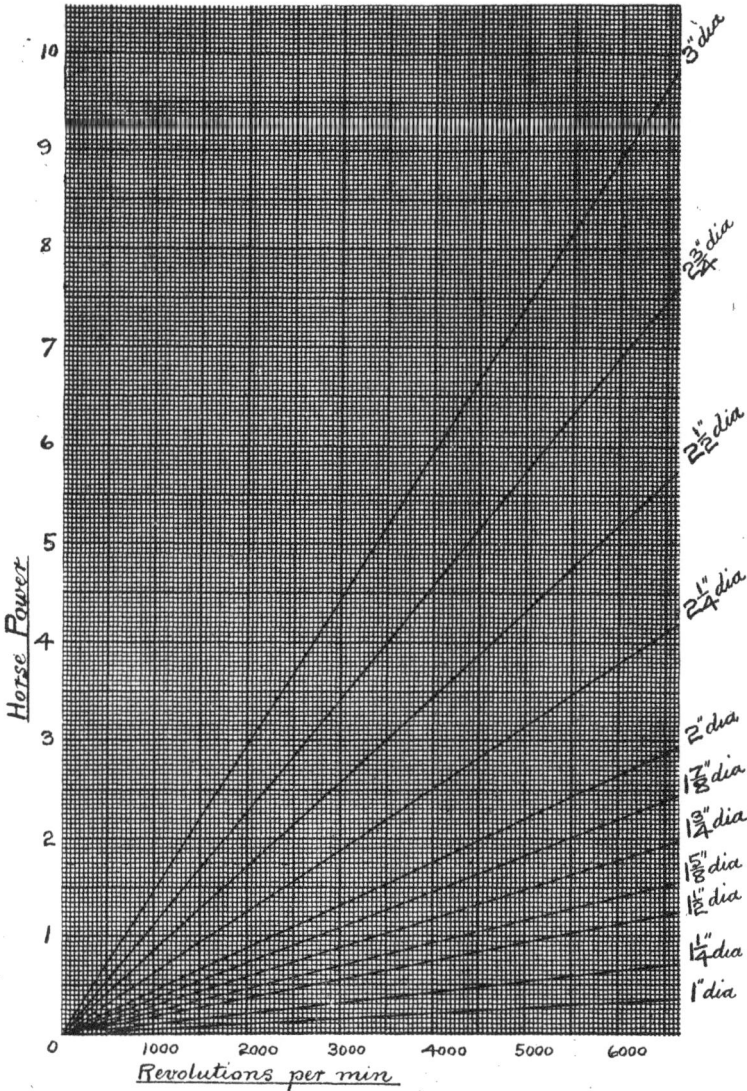

FIG. 28.—Horse-power absorbed by bearing friction.

4

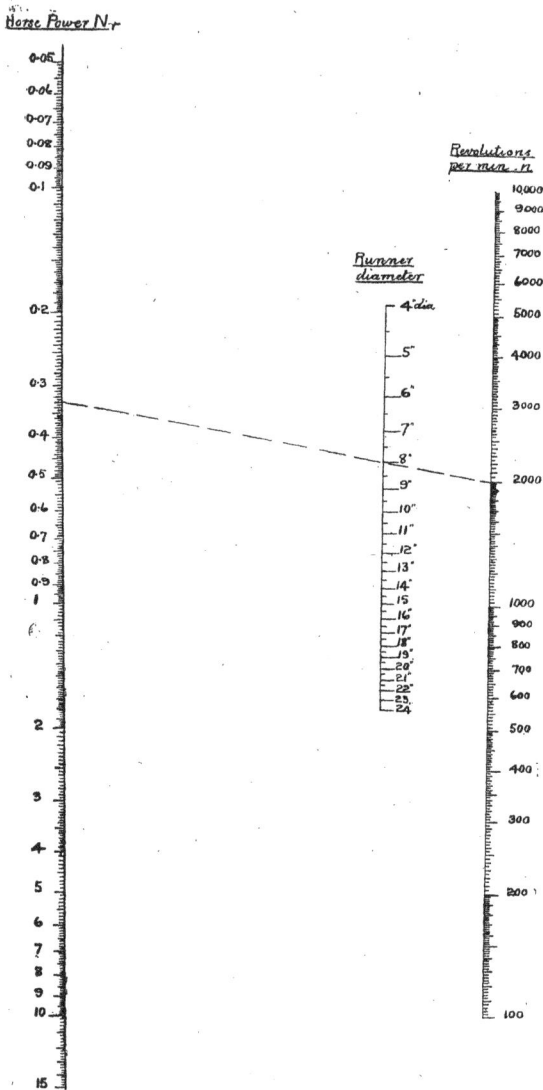

FIG. 29.—Horse-power absorbed in overcoming friction of runner sides.

Where d = diameter of journal and n the number of revolutions per minute.

In order to take into account the stuffing box friction the above value will be doubled and if N_b denote power required to overcome journal and stuffing box friction,

Then $N_b = 0.0000554d^3n$.

Fig. 28 represents curves with N_b as ordinates and n as abscissa for various diameters d of journals.

The power required to drive pump is, therefore, equal to the sum of the theoretical horse-power, the horse-power required to overcome runner friction (outside surfaces) and that necessary to overcome journal and stuffing box friction.

Denoting the power necessary to drive the pump by N, then

$$N = N + N_r + N_b.$$

If $N_w = \dfrac{H_m Q_a G}{550}$, or what is termed the water horse-power,

then if η_0 denotes the over-all efficiency,

$$\eta_0 = \frac{N_w}{N_t + N_r + N_b}.$$

CHAPTER VI.

EFFECT OF ANGLE AT DISCHARGE ON THE EFFICIENCY OF A CENTRIFUGAL PUMP.

THE vane angle β_2 at the periphery of the runner has a considerable influence on the hydraulic efficiency, the over-all efficiency of centrifugal pumps and the peripheral velocity required, also diameter of runner to pump against a given head.

The influence of the vane angle β_2 on the efficiency and velocity also depends on the means utilized to convert the kinetic energy at discharge from the runner into pressure energy, i.e. the type of casing, and whether guide vanes are used or not.

In order to show the effect of β_2, the three following cases are taken, and the peripheral velocity u_2, hydraulic efficiency η_h, and over-all efficiency η_e, are calculated for different angles β_2 when pumping 400 gallons per minute against a manometric head of 35 ft. at 900 revolutions per minute :—

1st Case.—With a badly designed volute, or where the whole of the kinetic energy at discharge from runner is lost.

2nd Case.—With a volute so designed that the gain in pressure head in volute is a maximum or $v_v = \dfrac{v_2 \cos \alpha_2}{2}$.

3rd Case.—With a ring of guide vanes outside the runner, the velocity of discharge from these guide vanes being equal to $\frac{1}{3}$ of the absolute velocity at inlet to guides.

In each case allowance has been made for the friction losses due to flow through runner and guide passages, but no allowance has been made for the friction losses in the casing or volute ; anyhow, these latter would be small compared with the former.

(52)

In cases (1) and (2) the friction losses l_2 and the losses l_1 and l_2 due to vane thicknesses are taken as $2g(l_1 + l_2 + l_3)$ = 216.

In case (3) these losses are taken as $2g(l_1 + l_2 + l_3 + l_4)$ = 262.

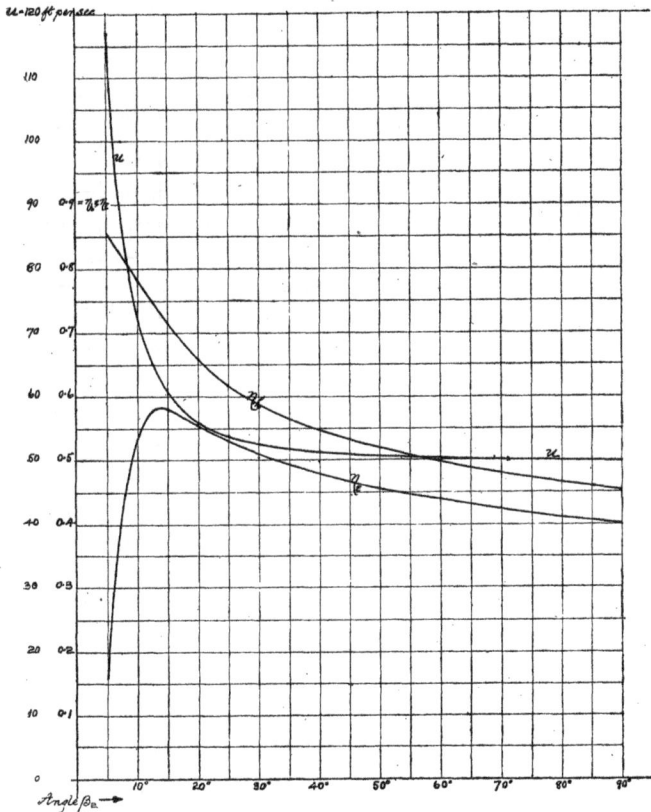

Pump no volute or badly designed one

FIG. 30.—Variation of peripheral velocity u_2, η_h, and η^e with angle β_2.

The radial component of relative velocity at entry to runner, $w_1 \sin \beta_1$, has been taken constant for all exit angles and equal to $0.2 \sqrt{2gh}$, or 9.5 ft. per second for cases (1) and (2).

The angle β_1 is equal to $21° \ 12'$, r_1 being constant for all values of β_2.

In the third case the radial component of relative velocity at inlet, $w_1 \sin \quad$, is taken as equal to $0·15 \ \sqrt{2gh} = 7·1$ ft.

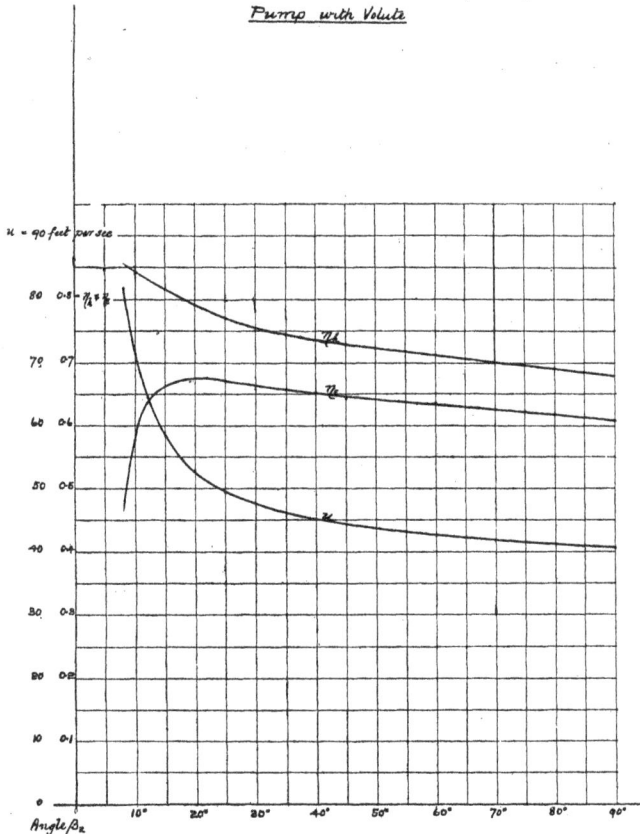

FIG. 31.—Variation of peripheral velocity u_2, η_h, and η_e with angle β_2.

(this agrees with practice in this type of pump), β_1 and r_1 being constant for all values of β_2.

The results for the above three cases are shown graphically in the curves in Figs. 30, 31, and 32, whose ordinates are u_2, η_h, and η_e, and abscissæ are the angles β_2.

Case I.—Fig. 30 shows that the hydraulic efficiency increases as the angle β_2 decreases, that is, the smaller the angle the greater the proportion of pressure energy produced in the runner.

Also to overcome a given head, the peripheral velocity is

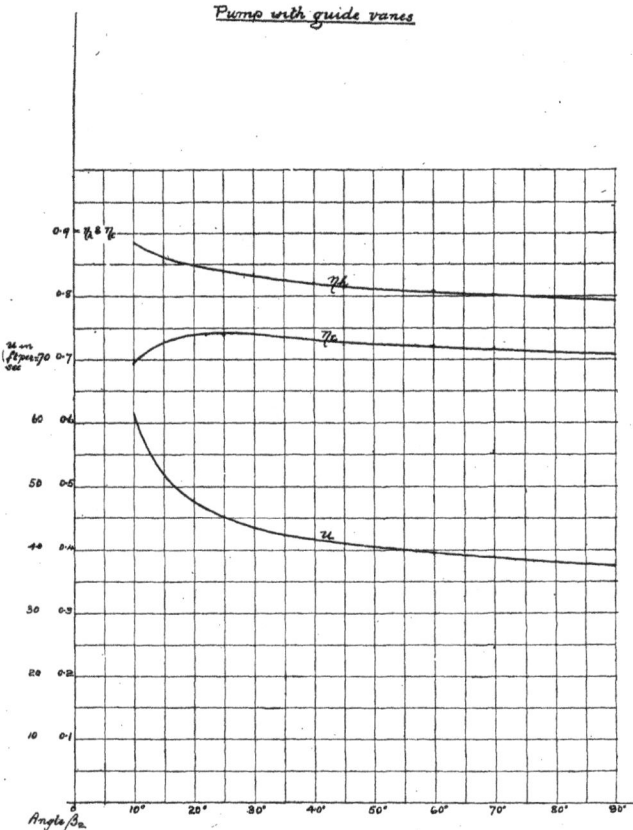

FIG. 32.—Variation of peripheral velocity u_2, η_h, and η_e with angle β_2.

greater the smaller the value of the angle β_2, which means that the diameter of runner is greater the smaller the value of β_2.

The efficiency η_e is a maximum for this type of pump when β_2 is in the neighbourhood of 14°, η_e decreasing as β_2 increases from 14° to 90°; η_e also decreases rapidly as β_2 decreases from 14°.

This rapid decrease in η_e (and increase in u_2) for decreasing values of β_2 below $14°$ is due to the increased power absorbed in overcoming the friction of the sides of the runner, this power varying as the fifth power of the diameter and the cube of the number of revolutions.

These curves show that for a pump of this type the best value of β_2 is about $14°$, the efficiency η_e being 58 per cent., the hydraulic efficiency η_h being $72·5$ per cent., and the peripheral velocity u_2 equal to $62·5$ ft. per second for this duty.

Case 2.—Fig. 31 represents the same quantities for a pump with a volute designed for each runner.

As in the preceding case, both the hydraulic efficiency η_h and the peripheral velocity u_2, increase as β_2 diminishes, but not so rapidly as in case 1.

The efficiency η_e has a maximum value of $67·5$ per cent. for $\beta_2 =$ about $21°$.

As the angle β_2 diminishes from $21°$, η_e diminishes, the rate of decrease of η_e with respect to β_2 increasing very rapidly for values of β_2 below $10°$; also as β_2 increases from $21°$, η_e decreases, but not so rapidly as in case 1 ; from $\beta_2 = 21°$ to $\beta_2 = 30°$ the decrease in efficiency is only about 1 per cent., whilst the peripheral velocity u_2 decreases from $51·5$ to $47·5$ ft. per second.

This means that the diameter of the runner can be decreased from $10·9$ ins. to $10·1$ ins., and the diameter of the casing decreased, by increasing the value of β_2 from $21°$ to $30°$, obtaining a lighter and cheaper pump with a sacrifice of 1 per cent. in the efficiency.

For this type of pump, taking all things into consideration, the value of β_2 may be made equal to $30°$, and this agrees closely with practice.

Case 3.—Fig. 32 shows the variation of $u_1\ \eta_h$ and η_e for pumps having guide vanes outside the runner.

Comparing these curves with those of cases 1 and 2, it will be seen that the η_e and η_h curves are much higher, also that the curve u is much lower; in other words, the pressure necessary for the lift of 35 ft. is obtained with much lower peripheral velocities, consequently with smaller diameter of runner when correctly designed vanes are fitted, the smaller diameter of runner considerably reducing the power absorbed by the friction of the outer surfaces of the rotating runner. Both η_h and u increase as β_2 diminishes.

The maximum value of η_e occurs when β_2 = about $25°$; when β_2 is less than $20°$, η_e begins to diminish rather rapidly its rate of diminution increasing as β_2 decreases.

As β_2 increases from $25°$ to $90°$, η_e decreases from $74\cdot5$ to 71 per cent., thus showing that with large values of β_2, say from $45°$ to $80°$, a good over-all efficiency can be obtained when properly designed guide vanes are used.

These results show how the efficiency varies with the angle β_2, also if β_2 is kept constant the efficiency is found to vary with the radial component at outlet from runner.

Therefore, for given type of pump, there is one value of the ratio x, that is, of the ratio of $w_2 \sin \beta_2$ to u_2, the peripheral velocity, that will give the best efficiency.

In order to obtain high efficiencies the width of the runner at outlet b_2 must never be made less than $0\cdot1r_2$. Should b_2 be less than $0\cdot1r_2$, then a higher speed of rotation giving a smaller value of r_2 should be used.

The following data may be used for the design of pumps having smooth passages :—

For pumps with no volute, or a badly-designed one, and with an angle at outlet from runner of about $12°$, the radial component $w_2 \sin \beta_2 = 0\cdot13u_2$.

$$\eta_h = 0\cdot62 \text{ in pumps where } b_2 = 0\cdot1r_2$$
$$\eta_h = 0\cdot72 \text{ ,, ,, ,, } b_2 = 0\cdot2r_2.$$

For pumps with a well-designed volute, and having values of β_2 from $22°$ to $30°$, the radial component should be $0\cdot18u_2$, and for $\beta_2 = 12°$ it should be $0\cdot09u_2$.

$$
\begin{aligned}
&\beta_2 = 12^\circ \\
&x = 0.09u_2
\end{aligned}
\begin{cases}
\eta_h = 0.7 & \text{for } b_2 = 0.075r_2 \\
\eta_h = 0.72 & ,, \quad ,, = 0.1r_2 \\
\eta_h = 0.76 & ,, \quad ,, = 0.15r_2 \\
\eta_h = 0.8 & ,, \quad ,, = 0.2r_2
\end{cases}
$$

$$
\begin{aligned}
&\beta_2 = 22^\circ \\
&x = 0.18u_2
\end{aligned}
\begin{cases}
\eta_h = 0.68 & ,, \quad ,, = 0.075r_2 \\
\eta_h = 0.7 & ,, \quad ,, = 0.1r_2 \\
\eta_h = 0.74 & ,, \quad ,, = 0.15r_2 \\
\eta_h = 0.78 & ,, \quad ,, = 0.2r_2
\end{cases}
$$

$$
\begin{aligned}
&\beta_2 = 30^\circ && \eta_h = 0.67 \quad ,, \quad ,, = 0.1r_2 \\
&x = 0.18x && \eta_h = 0.71 \quad ,, \quad ,, = 0.2r_2
\end{aligned}
$$

For all other values of x, the hydraulic efficiencies η_h will be less than those given.

In pumps with properly-designed guide vanes surrounding the runner, the following values should be used :—

$$
\begin{aligned}
&\text{For } \beta_2 = 12^\circ \ x = 0.06 \\
&,, \quad \beta_2 = 22^\circ \ x = 0.12 \text{ to } 0.18 \\
&,, \quad \beta_2 = 30^\circ \ x = 0.08 \text{ to } 0.14
\end{aligned}
$$

$$
\begin{aligned}
&\beta_2 = 12^\circ \\
&x = 0.06
\end{aligned}
\begin{cases}
\eta_h = 0.8 & \text{for } b_2 = 0.1r_2 \\
\eta_h = 0.85 & ,, \quad ,, = 0.2r_2.
\end{cases}
$$

$$
\begin{aligned}
&\beta_2 = 22^\circ \\
&x = 0.12 \text{ to } 0.18
\end{aligned}
\begin{cases}
\eta_h = 0.83 & ,, \quad ,, = 0.1r_2 \\
\eta_h = 0.88 & ,, \quad ,, = 0.2r_2.
\end{cases}
$$

$$
\begin{aligned}
&\beta_2 = 30^\circ \\
&x = 0.8 \text{ to } 0.14
\end{aligned}
\begin{cases}
\eta_h = 0.8 & ,, \quad ,, = 0.1r_2 \\
\eta_h = 0.83 & ,, \quad ,, = 0.2r_2.
\end{cases}
$$

For diameters of runners less than 8 ins. the values of η_h should be decreased.

CHAPTER VII.

DETAILS OF CENTRIFUGAL PUMPS.

THE materials of which centrifugal pumps are constructed depend upon the nature of the fluid to be pumped, the head against which pump works, and the speed.

The following list will give one some idea of the materials to be used :—

Liquid.	Specific Gravity.	Materials.
Acids . .	—	Casing and runner, acid-resisting bronze shaft, all bronze or steel bronze cased.
Beer and brewers' wort .	1·03	Cast-iron casing, gunmetal runner, tinned rolled bronze shaft.
Caustic potash and soda .	—	Cast-iron casing and runner, steel shaft ; no gunmetal or whitemetal in contact with shaft.
Ammoniacal liquor .	1·02	Cast-iron runner and casing, steel shaft with cast-iron sleeves, suction under pressure.
Petrol . .	—	Gunmetal casing and runner, steel shaft with gunmetal sleeves.
Benzine .	0·7	,, ,, ,,
Naphtha .	0·92	,, ,, ,,
Screened Sewage .	1·1	Cast-iron casing and runner (or gunmetal runner may be used).
Sea Water .	1·027	Admiralty practice : gunmetal casing and runner, bronze shaft. Ordinary practice : cast iron-case, gunmetal runner and guides, bronze cased shaft.
Beetroot juice .	1·1	Cast-iron case, cast-iron runner and guides.
Milk of lime	1·3	,, ,, ,, ,, ,,
Sugar syrup	1·3	Cast-iron case, gunmetal runner.

For pumps used in unwatering mines, where the water is slightly acid, the runners and guides are of gunmetal, and the shaft of nickel steel with gunmetal sleeves.

This is the usual practice with high-class multi-stage pumps working against high heads.

(59)

Runner.—The material of which the runner is constructed depends, as is seen from the above table, on the kind of fluid

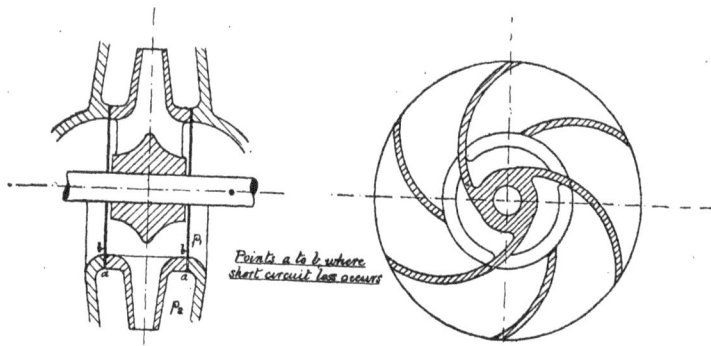

Points a to b where short circuit loss occurs

FIG. 33.—Runner (double inlet).

pumped, and also on the speed of the periphery of the runner ; this again limits the head attained per stage.

For cast-iron, the limiting value of u_2 may be taken as

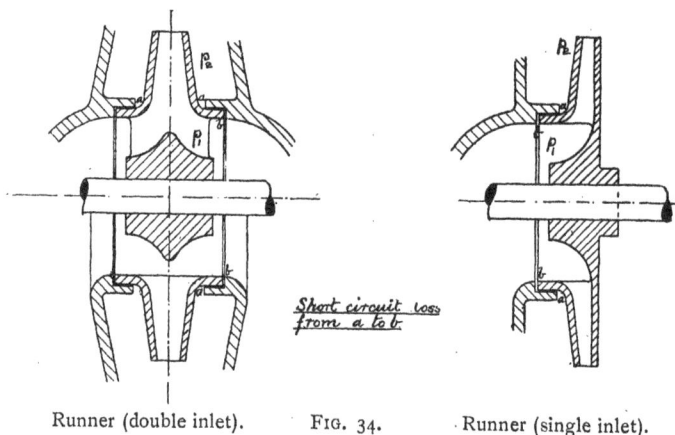

Short circuit loss from a to b.

Runner (double inlet).   FIG. 34.   Runner (single inlet).

100 ft. per second, for good gunmetal or bronze as 150 to 160 ft. per second.

The runner should be machined all over the outer surfaces and they should be finished quite smooth in order to reduce friction ; the passages should be finished as smooth as possible and polished (this latter remark applies equally to the guide passages in the diffuser).

The joint between the delivery side of runner and the inlet is formed by the runner and the casing a, b, Figs. 33, 34, and 35.

There must necessarily be some clearance between this part of the runner and the casing, but this clearance should be kept as small as possible ; its amount will depend on the stiffness of the shaft.

The greater this clearance the greater will be the quantity of fluid short-circuited which will be proportional to this

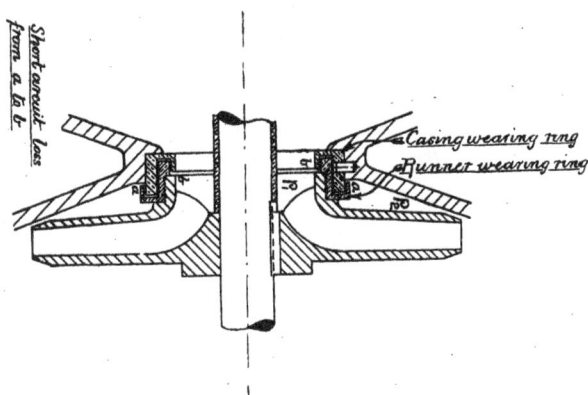

FIG. 35.—Single inlet runner with labyrinth joint.

clearance and to the square root of the pressure difference between the two sides of the joint a and b.

Figs. 33, 34 and 35 show sections of typical runners ; Fig. 35 illustrates a labyrinth type of joint designed by the American Well Works to reduce short-circuit losses, dissipating the head causing flow, by changing suddenly the direction of the flow of the fluid, and the areas past which it leaks.

Fig. 33 is the worst form, although it is still used by some makers.

Fig. 34 is a much better form, and quite as simple to construct and is used on most multi-stage pumps.

Fig. 35 is the best of the three types, especially when the

quantity of fluid is small compared with the head per stage against which it is to be pumped.

The runners should always be carefully balanced; they can be balanced statically with considerable accuracy, and this is usually sufficient to obtain good running.

With regard to the design of the vanes and passages, this will be treated in a future chapter dealing with the calculation and design of several pumps.

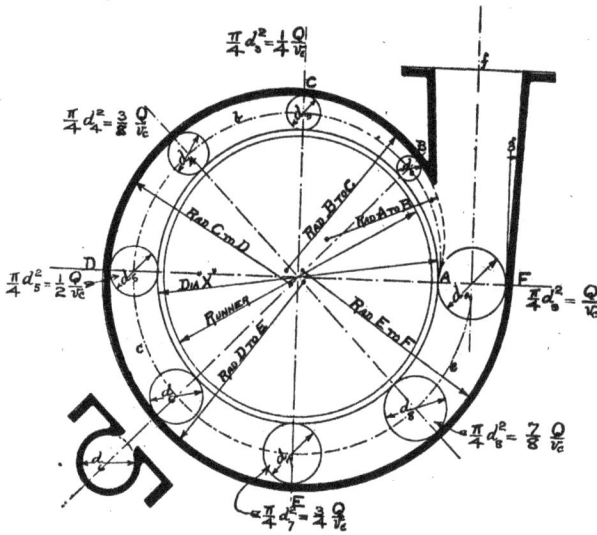

FIG. 36.—Spiral casing.

Casing.—For a single-stage pump, the only type of casing that can be considered, both from the points of view of efficiency and of weight, is the volute or spiral form with gradually increasing cross section.

In single-stage pumps with guides, as shown in Fig. 24, and for the last stage of multi-stage pumps, a cylindrical form is often used; the velocity, being wholly lost by shock, is kept smaller than would be the case in a volute.

Dealing with the volute form with circular cross sections, as shown in Fig. 36, the cross section at AF is made to carry the whole of the fluid dealt with by the pump, when flowing past the section named with the velocity v_c.

The velocity v_c being constant, the areas of the cross sections are made proportional to the amount of fluid passing them.

For instance, past the section at D, half of the fluid dealt with by the pump passes, or $\frac{1}{2}Q$, therefore the area at D will be half the area at AF, and the area at E will be $\frac{3}{4}$ the area at AF.

Fig. 36 shows this, and also the method of setting out the boundary curve BCDEF.

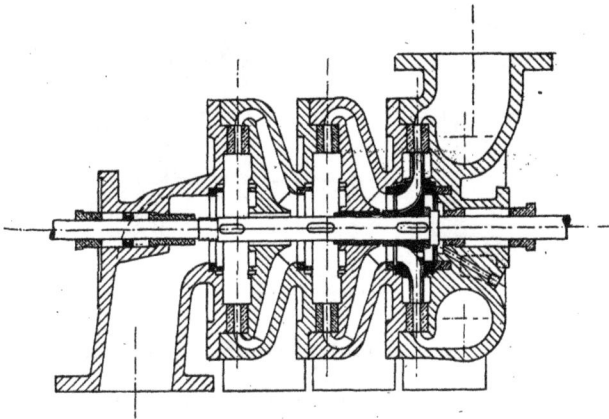

FIG. 37.—Casing for multi-stage pump built up of sections.

In some cases the section of the volute is made rectangular, also at times the discharge pipe changes from rectangular section AF to a circular section at discharge f.·

If with a volute of rectangular cross section the breadth is constant, then the depth from the base circle of diameter X, to the boundary curve BCDEF will be directly proportional to the quantity passing.

The casings for multi-stage pumps are illustrated in Figs. 37 and 38 in diagrammatic form and are typical of the two methods used by builders.

Shaft.—The material of the shaft depends on the nature

of fluid to be pumped, and its dimensions on the power passing and the weight of the runners.

Where possible, the shaft should be made of nickel steel and should be fitted with renewable gunmetal sleeves where it passes through glands and stuffing boxes.

The shaft is subjected to a twisting moment together with a bending moment produced by the weight of the runners.

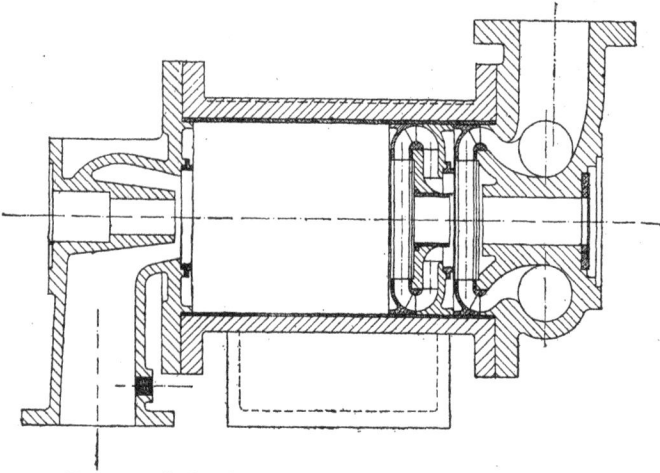

FIG. 38.—Casing for multi-stage pump. Cylindrical type.

The twisting moment T to which the shaft is subjected is, if

N = power required to drive the pump,

n = number of revolutions per minute,

$$T = \frac{N \times 33,000 \times 12}{2\pi n} \text{ lb. ins.}$$

The resistance of a shaft to torsion is given by the formula

$$\frac{\pi}{16} d^3 f$$

and this must equal T,

where d = diameter of shaft in inches,

f = shear stress allowable at outer fibres of the shaft, f being a maximum at the surface of shaft.

The smallest dimension of the shaft, which is in the coupling, is calculated from the above equations.

f may be taken as from 8000 to 10,000 lb. sq. ins.

From this point the shaft is made larger in diameter in the bearings, and again larger in diameter in the runners. Fig. 39

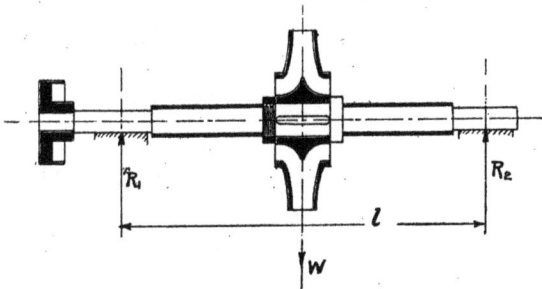

Fig. 39.—Shaft for single-stage pump.

shows a shaft for a single-stage pump, and Fig. 40 shows that of a multi-stage pump.

At a the shaft is in pure torsion; in the centre directly under the line of action of the load due to the weight of the

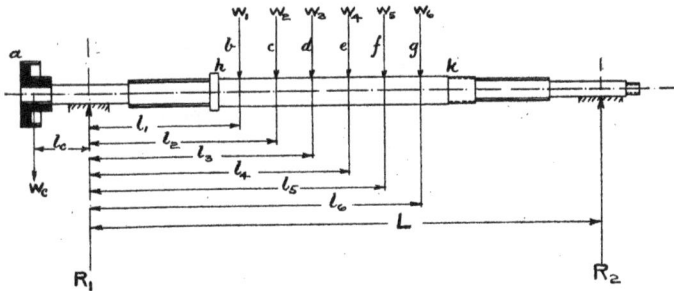

Fig. 40.—Shaft for multi-stage pump.

runner W, the maximum bending moment M due to this load occurs, Fig. 39.

At the centre, then, the shaft is subjected to a bending moment M together with a twisting moment T.

If the span of the shaft be taken as the distance between the centres of the bearings and is l,

5

The maximum bending moment at the centre is

$$M = \frac{Wl}{4}.$$

The twisting moment is, as before,

$$T = \frac{N \times 33,000 \times 12}{2\pi n}.$$

Using the "Guest" formula for combined twisting and bending,

$$T_e = \sqrt{M^2 + T^2},$$

T_e being the equivalent twisting moment.

The diameter of the shaft is then obtained from the equation

$$T_e = \frac{\pi}{16} f d^3,$$

d here being the diameter where the corresponding value of M is taken to cause the maximum shear stress f.

Fig. 40 represents a shaft for a six-stage pump. In this the twisting moment varies along the shaft—from a to b it is T lb. ins., from b to c it is $\frac{5}{6}$T, and, finally, from f to g it is $\frac{1}{6}$T.

The next operation is to find the bending moments along the shaft.

$$R_2 = \frac{W_1 l_1 + W_2 l_2 + W_3 l_3 + W_4 l_4 + W_5 l_5 + W_6 l_6 - W_c l_c}{L}.$$

W_1, W_2, W_3, etc., are the weights of the runners, and are equal. W_c is the weight of the coupling

$$R_1 = W_1 + W_2 + W_3 + W_4 + W_5 + W_6 + W_c - R_2.$$

The bending moment at different points along the shaft are :—

$$M_b = R_1 l_1 - W_c(l_1 + l_c),$$
$$M_c = R_1 l_2 - W_c(l_2 + l_c) - W_1(l_2 - l_1),$$
$$M_d = R_1 l_3 - W_c(l_3 + l_c) - W_1(l_3 - l_1) - W_2(l_3 - l_2),$$
$$M_e = R_1 l_4 - W_c(l_4 + l_c) - W_1(l_4 - l_1) - W_2(l_4 - l_2) - W_3(l_4 - l_3),$$
$$M_f = R_1 l_4 - W_c(l_5 + l_c) - W_1(l_5 - l_1) - W_2(l_5 - l_2) - W_3(l_5 - l_3)$$
$$- W_4(l_5 - l_4),$$
$$M_g = R_1 l_5 - W_c(l_6 + l_c) - W_1(l_6 - l_1) - W_2(l_6 - l_2) - W_3(l_6 - l_3)$$
$$- W_4(l_6 - l_4) - W_5(l_6 - l_5).$$

The equivalent twisting moments are :—

$$\sqrt{M_b{}^2 + T^2} \text{ at } b, \ \sqrt{M_c{}^2 + (\tfrac{5}{6}T)^2} \text{ at } c, \text{ and so on.}$$

The maximum equivalent twisting moment is found, and from it, and the allowable stress, the diameter is calculated, the shaft being made this diameter from h to k. At a in the coupling the shaft is calculated for twisting only.

The diameter of the shaft is increased in the first bearing

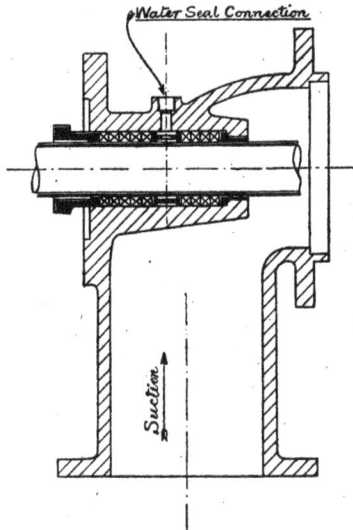

FIG. 41.—Suction branch.

by about $\frac{1}{4}$ inch larger than at a, and both bearings are usually made the same for practical reasons.

The dimensions of the shaft must be such that the critical speed is at least 30 to 40 per cent. above the running speed, and if this is not the case the dimensions must be increased until the critical speed is raised to this margin above the running speed.

For the case in Fig. 39, the critical speed may be as

$$n_c = a\sqrt{\frac{1}{W_c l^3}}.$$

n_c is the critical speed in revolutions per minute.

W_c is combined weight of runner and shaft.

I is the moment of inertia of the shaft.

a is a constant which may be taken as about 420,000 for flexible bearings, and about 840,000 for rigid bearings.

For the more complicated cases, as in Fig. 40, the reader is referred to larger treatises ("Steam Turbine" Stodola).

Stuffing Boxes and Glands.

Great care should be exercised in the design of the stuffing boxes and packings.

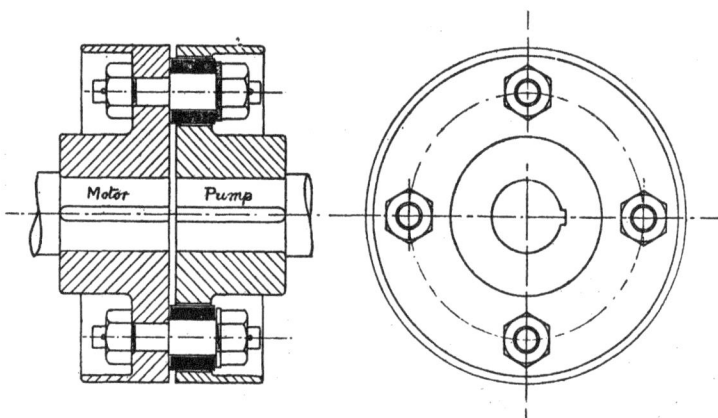

Fig. 42.—Flexible shaft coupling.

At the suction end the stuffing box and packing is usually to prevent the admission of air into the pump; and at the delivery end to prevent leakage out of the pump.

Fig. 41 shows a design of stuffing box, with a lantern bush and connection which is connected to the first or second stage of the pump, in the case of the suction end.

The pressure water forms a water seal which effectively prevents the admission of air into the suction, which is usually less than atmospheric pressure.

At the delivery end the lantern bushes are connected to a drain or point of lower pressure.

The stuffing-boxes should be long and able to hold at least six to eight turns of soft packing in addition to the lantern bushes, in order that the glands need not be tightened too much.

The coupling through which the pump is driven by the motor should be of the flexible type.

Fig. 42 shows a very satisfactory type, in which pins, tight in one half, drive the other by means of rubber bushings.

Great care should be taken that the pump and its motor are in alignment, because the coupling illustrated is constructed so that any end play of the motor shaft will not affect the pump ; allow pump spindle to take up any position determined by the action of the balancing device.

CHAPTER VIII.

AXIAL THRUST AND ITS BALANCING.

IN single-stage pumps the entrance to the runner is on both sides, unless the quantity is so small that entry can only be on one side in order to obtain a reasonable diameter of eye.

In the case of a runner having the inlet on both sides, the total axial forces on the sides of the runner are equal, and opposite in direction, therefore there is no axial or end thrust, providing the sides of the pump casing are at equal distances from the runner sides.

In high pressure pumps, where a number of stages are necessary to obtain the desired head, and the quantity delivered is relatively small, the runner having a single inlet on one side is the only one that can be used.

The eyes of the double inlet would be small, and would necessitate very complicated passages in the casing, offsetting the gain obtained by having the runners in axial balance.

Therefore, for multi-stage pumps runners having a single inlet are the rule, the passages from one stage to the next being of simple construction and end or axial thrust being taken up by some form of balancing arrangement.

In the following discussion forces acting from right to left will be considered as positive, and those acting from left to right will be, therefore, negative.

Considering a pump (or one stage of a multi-stage pump), as Fig. 43, the runner having a single inlet, the pressure at inlet being p_i, and the pressure on the outer faces of the runner being p_o :—

The pressure p_o is not constant; it varies with the radius owing to the rotation of the water in spaces between sides.

(70)

The pressure p_0 will be taken as that of the outlet from runner and will also be considered as constant (that is, variation in p_0 due to rotating paraboloid will be neglected) over outer surfaces of runner.

F_r = total force acting from right to left.

F_l = ,, ,, ,, ,, left to right.

Then $T = F_r - F_l$,

T being the end thrust which has to be provided for.

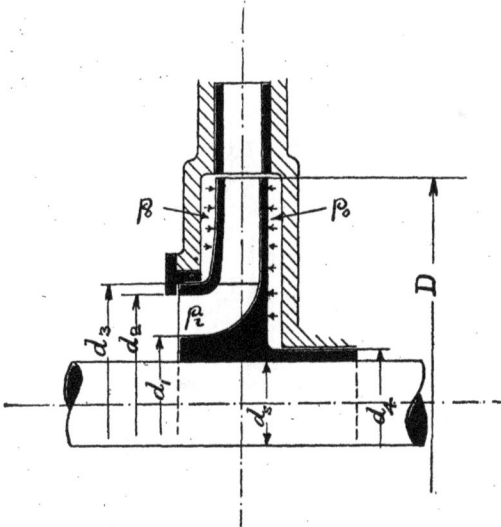

FIG. 43.—Diagram of runner showing pressures which produce end thrust.

$$F_r = \frac{\pi}{4}(D^2 - d_4^2)p_0$$

$$F_l = \frac{\pi}{4}(D^2 - d_3^2)p_0 + \frac{\pi}{4}(d_3^2 - d_1^2)p_i + \frac{\pi}{4}(d_2^2 - d_1^2)\frac{v^2}{g}.$$

The last term, namely, $\frac{\pi}{4}(d_2^2 - d_1^2)\frac{v^2}{g}$, is a force due to change of momentum due to velocity of entering fluid.

$$T = \frac{\pi}{4}\left\{(D^2 - d_4^2)p_0 - (D^2 - d_3^2)p_0 - (d_3^2 - d_1^2)p_i - (d_2^2 - d_1^2)\frac{v^2}{g}\right\}$$

$$= \frac{\pi}{4}\left\{(d_3^2 - d_4^2)p_0 - (d_3^2 - d_1^2)p_i - (d_2^2 - d_1^2)\frac{v^2}{g}\right\}.$$

If, for the time being, the term $(d_2{}^2 - d_1{}^2)\dfrac{v^2}{g}$ be neglected, and also if $d_1 = d_4$, then,

$$T = \frac{\pi}{4}\{(d_3{}^2 - d_4{}^2)p_o - (d_3{}^2 - d_4{}^2)p_i\}$$

$$= \frac{\pi}{4}\{(d_3{}^2 - d_4{}^2)(p_o - p_i)\}.$$

If another joint be made on the back of the runner, as in Fig. 44, of diameter equal to d_3, and making the pressure in

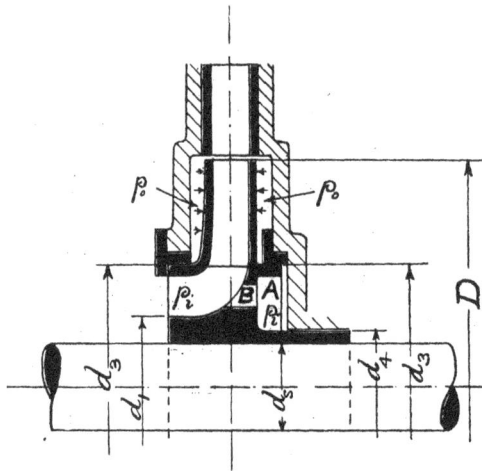

FIG. 44.—Diagram of runner showing method of eliminating end thrust.

this space of diameter d_3 at the back of the runner equal to p_i the sum of the axial forces due to pressure only will be zero, that is $T = 0$.

The equalizing of the pressure in this space A with inlet pressure p_i is obtained by connecting this space, by means of a pipe, with the runner inlet, or by holes B in the runner itself.

In order to allow for the thrust due to momentum the diameter of joint at back of the runner should be slightly greater than d_3; this would give a balance for one given

quantity, but with any change in the quantity delivered this
balance would immediately
be upset owing to the
change in v, the thrust due
to momentum varying as
v^2, which is, as Q^2.

This method of reduc-
ing end thrust has been
used by certain makers,
notably Jaeger's, and Wor-
thingtons.

In this type of impeller
there is more than twice
the short-circuit loss, first
due to the extra joint,
secondly the lowering of
the pressure in space A,
which increases the leak-
age or short-circuit loss
from the stage above.

From these considera-
tions it is evident that
runners of the type in Fig.
43 are the best if some
external arrangement can
be devised to neutralize
the end thrust. This is
now done by means of a
piston subjected to fluid
pressure equal to, and acting in a direction opposite
to, the resultant axial out
of balance force of the
runner or runners.

A multi-stage pump
having runners as in Fig. 43 is shown in Fig. 45, a three-

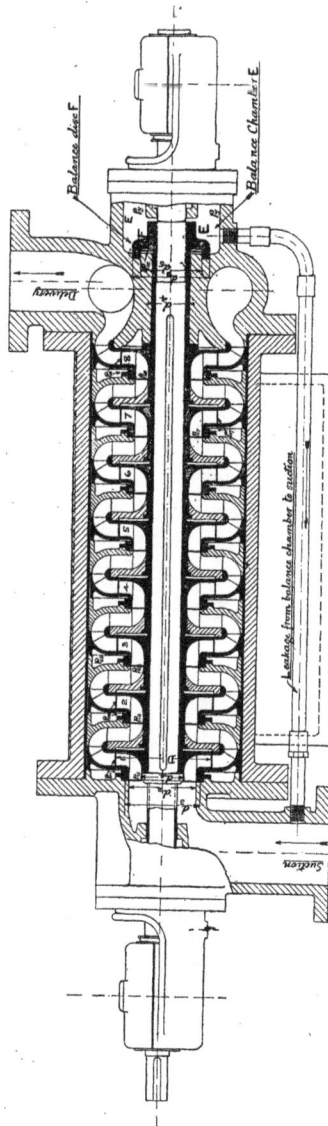

FIG. 45.—Section of 8-stage pump showing hydraulic balancing device.

stage pump being taken as an example, the reasoning apply-ing to any number of stages.

Velocity v is common to all stages; the total force due to momentum is

$$3\frac{\pi}{4}(d_2{}^2 - d_1{}^2)\frac{v^2}{g} = 3\frac{\pi}{4}(d_2{}^2 - d_4{}^2)\frac{v^2}{g}$$

since $d_4 = d_1$ and acts in a direction from left to right.

Unbalanced forces due to pressure only acting right to left for the different stages are as follows :—

Runner No. 1 $\dfrac{\pi}{4}(d_3{}^2 - d_4{}^2)\,(p_{o_1} - p_{i_1})$.

„   No. 2 $\dfrac{\pi}{4}(d_3{}^2 - d_4{}^2)\,(p_{o_2} - p_{i_2})$.

„   No. 3 $\dfrac{\pi}{4}(d_3{}^2 - d_4{}^2)\,(p_{o_3} - p_{i_3})$.

Resultant end thrust T is from right to left or towards suction end of pump.

$$T = \frac{\pi}{4}(d_3{}^2 - d_4{}^2)\,(p_{o_1} - p_{i_1}) + \frac{\pi}{4}(d_3{}^2 - d_4{}^2)\,(p_{o_2} - p_{i_2})$$

$$+ \frac{\pi}{4}(d_3{}^2 - d_4{}^2)\,(p_{o_3} - p_{i_3}) - 3\frac{\pi}{4}(d_2{}^2 - d_4{}^2)\frac{v^2}{g}.$$

$$= \frac{\pi}{4}(d_3{}^2 - d_4{}^2)3(p_{o_1} - p_{i_1}) - \frac{3\pi}{4}(d_2{}^2 - d_4{}^2)\frac{v^2}{2g}$$

since $(p_{o1} - p_{i1}) = (p_{o2} - p_{i2}) = (p_{o3} - p_{i3}) = (p_{on} - p_{in})$.

In the case of a pump with n runners,

$$T = \frac{\pi}{4}(d_3{}^2 - d_4{}^2)n(p_{oi} - p_{i1}) + n\frac{\pi}{4}(d_2{}^2 - d_4{}^2)\frac{v^2}{2g}.$$

The method adopted by the best makers to counteract the influence of this end thrust is, to balance it by the high pres-sure fluid from last runner or diffuser acting on a disc or piston F revolving in a chamber E. The chamber E has a large drain to carry away the leakage water either to the atmosphere and thence to the suction, or this chamber is connected direct to the suction-inlet of pump. Of these two methods, flow of leakage water to atmosphere is the best, as this leakage is seen, and the satisfactory working of the piston. The pressure in the chamber will therefore be just a little above the atmosphere, and for calculating the dimensions of the piston atmospheric pressure may be assumed.

Then
$$T = p_x \frac{\pi}{4} d_5^2 - p_y \frac{\pi}{4} d_6^2,$$

p_x = pressure on diameter d_5, which is due to that of last runner or diffuser.

p_y = pressure in chamber E.

The tendency of the disc is to move a certain distance to the right according to the value p_x acting on it, where it will remain in equilibrium; any further movement to right will be resisted and will cause the value of $p_y \frac{\pi}{4} d_6^2$ to be greater than

$$p_x \frac{\pi}{4} d_5^2.$$

The flow of water past the plate or leakage will be proportional to

$$\pi d_6 \times c \times \sqrt{p_x}$$

where c is clearance between working face of piston and its corresponding seat on chamber E.

In the radial type movement of this piston must be restricted to a small amount, otherwise the runners will not discharge directly into the guide or diffuser rings.

Therefore a light thrust bearing is necessary in these pumps in order to locate the runners with regard to the diffusers.

In the Kugel and Gelpke pump made by Messrs. Hayward Tyler & Co., Ltd., London, and Messrs. Escher Wyss & Co., Zurich, a thrust block is quite unnecessary, as whatever position rotor takes up longitudinally, discharge is always direct into guides.

Another point about this design of pump is that the thrust due to momentum is practically zero; the change of momentum at inlet is equal to change of momentum at outlet of runner but of opposite sign.

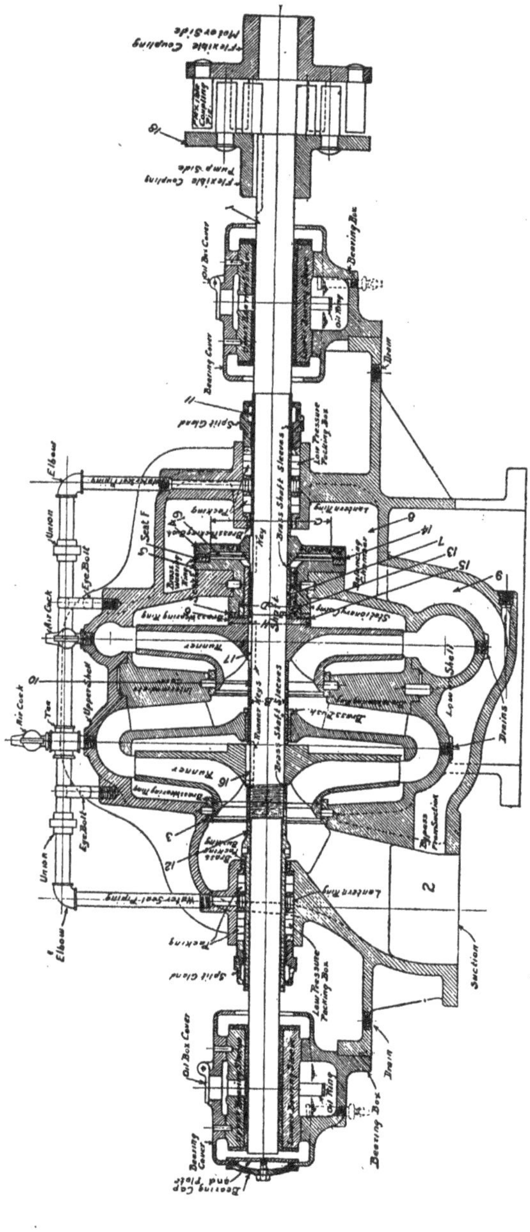

Fig. 46.—Section of 2-stage pump showing hydraulic balancing device as used by the American Well Works.

The sectional view Fig. 46 represents the balancing device as used by The American Well Works, Aurora, Ill., on a two-stage pump made by them.

The runners are mounted on the shaft I and driven by feathers 16 and 17. They are held in correct position lengthwise by shaft sleeves 11, 7, 10 and locknut 12.

The diameters of the opening at A are equal on both impellers and the diameters of the sleeves B are also equal, and since the area of the opening at A is greater than the area of diameter B, there is a tendency of the shaft I with the attached runners to move towards the suction opening 2 of the pump, as shown in the previous reasoning. To counteract this movement of the shaft and runners towards the suction opening a hydraulic balancing device is used.

The balancing mechanism consists of the stationary casting 5, with two removable bronze seats E and F, and rotating discs 4 and 6. (Disc 6 can be also the hub of the impeller.)

The balancing device is so designed that the difference in areas between disc 4, represented by diameter C, and disc 6, represented by diameter B, is sufficiently larger than the area represented by the opening A to make the shaft I move towards the flexible coupling in case the fluid enters the balancing device between the faces of the disc 6 and seat E and port 13.

However, the instant the shaft moves towards the motor the disc 6 is carried close to the seat E, preventing more fluid entering the balancing device.

This action carries disc 4 away from seat F and permits the fluid in spaces 13 and 19 to pass out into the balancing chamber 8, from which it flows by bye-pass 9 into the suction opening 2 of the pump.

The instant disc 6 comes in contact with seat E no more fluid can come into the spaces 13 and 19 and therefore the force for moving the shaft towards the flexible coupling has been entirely cut off, leaving the pump with a pressure toward the suction opening equal to the difference in areas between A and H multiplied by the pressure of the last stage.

The instant that the disc 6 moves away from seat E water will enter the balancing device through passages 15 and 13 into chamber 19, and cause the shaft I with all its moving parts to instantly move towards the flexible coupling.

This movement can be noticed in the first few revolutions of the pump when starting. It will, however, almost instantly find its equilibrium and no movement lengthwise can be detected when the pump is running, even with the assistance of trams.

An impression might be that these discs wear out quickly, but a careful examination of several which have had a large amount of service showed no wear whatever; in fact, the tool marks were still there.

This is due to the fact that even when either rotating disc is carried closest to a stationary disc the two are still separated by a film of fluid. In order to guard against possible wear casting 5 is provided with removable bronze seats E and F, which can be readily removed and others inserted, making the device as good as new.

CHAPTER IX.

CALCULATION AND DESIGN OF A SINGLE-STAGE PUMP.

To illustrate the principles of the previous chapters, applied to the design of centrifugal pumps, two cases and types will be taken and all necessary calculations made.

Case I.—A volute type of centrifugal to be capable of pumping 1000 gallons of water per minute against a total head of 60 ft.; including all frictional losses in suction and delivery pipes, and to be driven by an electric motor direct coupled to pump shaft, running at 1200 revolutions per minute :—

If Q be the actual quantity pumped in cubic feet per second then $Q = \dfrac{1000}{60 \times 6\cdot24} = 2\cdot67$ cub. ft. per second.

The quantity passing through the runner must be greater than this, by the amount of water short-circuited.

Calling Q_t the amount of water to be passed by the runner,

$Q_t = Q +$ amount of short circuit of leakage.

Q_t will be assumed equal to $2\cdot94$ cub. ft. per second.

$H_m = 60$ ft.

$\omega =$ angular velocity of runner $= \dfrac{2\pi n}{60} = 125\cdot7$.

$\beta_2 = 22°$, $\sin \beta_2 = 0\cdot3746$, $\cos \beta_2 = 0\cdot9272$, $\cot \beta_2 = 2\cdot4751$. $x = 0\cdot18$.

The hydraulic efficiency will at present be assumed as $0\cdot7$, which should be attained with the above values of β_2 and x, if the width of the runner at outlet is not less than $0\cdot1 r_2$.

This gives, that H_t, the work per lb. of water furnished by the runner, must be

$$H_t = \frac{H_m}{\eta_h} = \frac{60}{0\cdot7} = 85\cdot7.$$

$$(79)$$

$$H_t = \frac{u_2^{\,2}}{g}(1 - \cot \beta_2 x)$$

$$u_2 = \sqrt{\frac{gH_t}{1 - \cot \beta_2 x}} = \sqrt{\frac{32 \cdot 2 \times 85 \cdot 7}{1 - 0 \cdot 18 \times 2 \cdot 4751}} = \sqrt{\frac{2760}{0 \cdot 554}}$$

$$u_2 = \sqrt{4980} = 70 \cdot 5 \text{ ft./sec.}$$

$$w_2 \sin \beta_2 = 0 \cdot 18 u_2 = 12 \cdot 7 \text{ ft./sec.}$$

$$w_2 = \frac{12 \cdot 7}{\sin \beta_2} = 33 \cdot 9 \text{ ft./sec.}$$

$$r_2 = \frac{u_2}{\omega} = 0 \cdot 562, \ D_2 = \text{say } 13 \cdot 5 \text{ ins.}$$

Assuming that

$$\frac{p_2}{p_2 + t_2} = 0 \cdot 9$$

$$Q_t = 2\pi r_2 b_2 \times 0 \cdot 9 \times x u_2$$

$$b_2 = \frac{Q_t}{2\pi r_2 x u_2 \times 0 \cdot 9} = \frac{2 \cdot 94}{6 \cdot 28 \times 0 \cdot 562 \times 12 \cdot 7 \times 0 \cdot 9} = 0 \cdot 0728 \text{ ft.} = 0 \cdot 875'$$

the ratio of $\dfrac{b_2}{r_2} = \dfrac{0 \cdot 0728}{0 \cdot 562} = 0 \cdot 129$; this justifies the assumption that $\eta_h = 0 \cdot 7$.

To determine inlet dimensions, the shaft is assumed to have a diameter of $1\frac{1}{2}$ in. in the bearings, and 2 ins. in the middle where the runner is carried.

If the thickness of runner boss is taken as $\frac{1}{2}$ in. the diameter will be 3 ins.

Since inlet takes place on both sides of the runner, each inlet must pass $\dfrac{Q_t}{2}$ cub. ft. of fluid per second at a velocity equal to about $1 \cdot 3$ to $1 \cdot 5 x$, which in this case will be taken as $17 \cdot 8$ ft./sec.

$$\text{Annular area at entrance} = \frac{Q_t}{2 \times 17 \cdot 8} = \frac{2 \cdot 94}{2 \times 17 \cdot 8}$$
$$= 0 \cdot 0825 \text{ sq. ft.}$$
$$= 11 \cdot 9 \text{ sq. ins.}$$

$$11 \cdot 9 = \frac{\pi}{4} d_e^2 - \frac{\pi}{4} 3^2$$

$$\frac{\pi}{4} d_e^2 = 11 \cdot 9 + 7 \cdot 06 = 18 \cdot 96$$

$$d_e = 4 \cdot 9 \text{ ins., say } 5 \text{ ins.}$$

If d_e be taken as 5 ins. the velocity will be decreased from 17·8 to 16·9 ft. per second.

The radius r_1 of inlet edge of vanes will be taken as 2·75 ins. or 0·229 ft.

$$u_1 = \omega r_1 = 0 \cdot 229 \times 125 \cdot 7 = 28 \cdot 8 \text{ ft. per second.}$$

If $v_1 = 16 \cdot 9$ ft. per second and as a_1 has already been assumed as 90°, the angle of the vane at inlet β_1 may be determined.

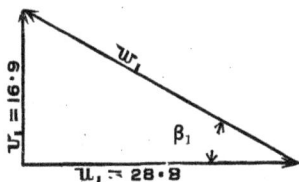

$$\beta_1 = \tan^{-1} \frac{16 \cdot 9}{28 \cdot 8} = \tan^{-1} 0 \cdot 587$$

$$\beta_1 = \text{say } 30° \ 30'.$$

If $\dfrac{p_1}{p_1 + t_1} = 0 \cdot 85$

breadth of runner at inlet may now be calculated

$$b_1 = \frac{Q}{2 \pi r_1 w_1 \sin \beta_1 \times 0 \cdot 85} = \frac{2 \cdot 94}{6 \cdot 28 \times 0 \cdot 229 \times 16 \cdot 9 \times 0 \cdot 85} = 0 \cdot 142 \text{ ft.}$$

$b_1 = 0 \cdot 142$ or 1·705 ins.

Design of the Runner.—The first thing to do is fix upon the profile of the runner, making the sides parallel for a short distance, as this is favourable for the discharge into the casing or guides when used.

The profile is shown in the left hand view on Fig. 47. The difference between r_1 and r_2 is divided into a number of equal parts; these are numbered from 1 to 9 in this example.

The peripheral velocity u_2 is set out to scale, from OY, and by joining c with O the linear velocity corresponding to different radii can be scaled off.

Assuming that the energy given to the fluid is proportional

6

to the radius, then knowing the radial component of velocity from the widths of the runner at different radii, and knowing u the velocity, diagrams giving w or the relative velocity along the vane can be drawn.

The directions of w are the tangents to the working face of the vane at different points.

The vane can now be drawn by means of circular arcs which will best suit the curve from a to b in Fig. 47.

In finding the values of w due allowance must be made for thickness of the vanes.

$\dfrac{p}{p+t}$ has been taken in Fig. 47 as 0·85 at inlet r_1 and 0·9 at outlet r_2.

The number of vanes varies somewhat with the value of β_2, and with the diameter of the runner; in this example the number of vanes is eight.

At inlet the vanes should be sharpened on the back, and also at the outlet, as shown in the figure; this is done to reduce the losses due to vane thickness at entry and outlet.

Casing.—The casing is designed as indicated in chapter VII. The area at AF is made to pass 2·67 cub. ft. per second at a velocity v_c equal to $\dfrac{v_2 \cos a_2}{2}$ which will be taken as 19 ft. per second.

Area at entrance to discharge pipe $= \dfrac{2·67}{19} = 0·14$ sq. ft.

$$a = 20·2 \text{ sq. ins.}$$

Making the cross sections of the volute circular, the diameter corresponding to $a = 20·2$ is 5·07 sq. ins., say $5\frac{1}{8}$ ins., this gives an area of 20·6 sq. ins.

Section.	Area.		Corresponding Diameter.
1	$a = 20·6$	sq. ins.	$5\frac{1}{8}$ ins.
2	$\frac{7}{8}a = 18·1$	,,	$4\frac{13}{16}$,,
3	$\frac{3}{4}a = 15·5$	,,	$4\frac{7}{16}$,,
4	$\frac{5}{8}a = 12·9$	,,	$3\frac{15}{16}$,,
5	$\frac{1}{2}a = 10·32$	,,	$3\frac{1}{2}$,,
6	$\frac{3}{8}a = 7·74$	,,	$3\frac{1}{8}$,,
7	$\frac{1}{4}a = 5·16$	,,	$2\frac{9}{16}$,,
8	$\frac{1}{8}a = 2·58$	,,	$1\frac{13}{16}$,,

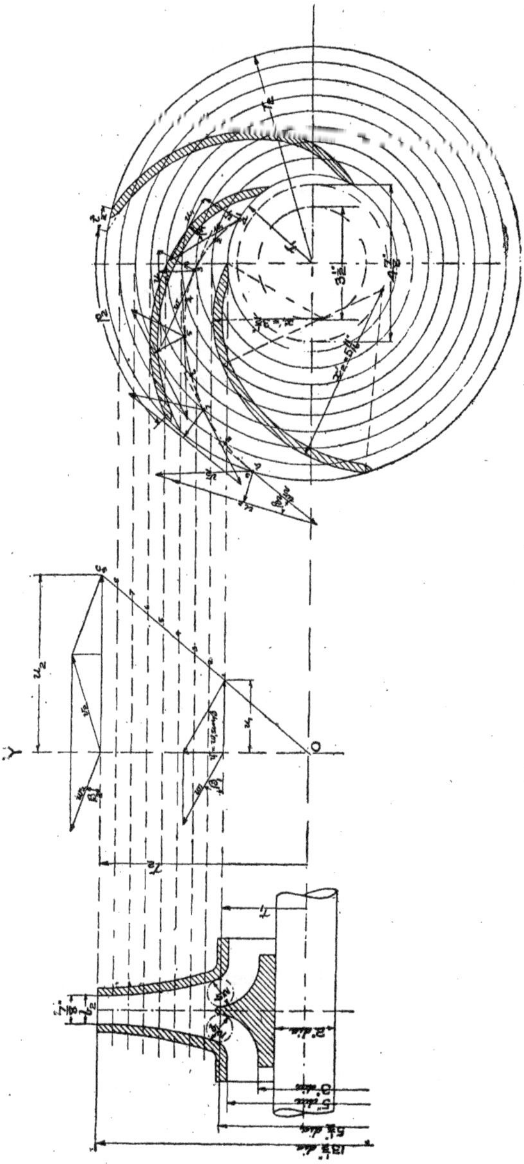

Fig. 47.—Diagram showing method of designing runner vanes.

Taking the velocity in the suction pipe as 10 ft. per second and the same for the discharge outlet, their diameters will be 7 ins., this gives a velocity equal to $0\cdot161\sqrt{2gh_m}$, which is good practice for this class of pump.

The discharge should therefore taper from $5\frac{1}{8}$ ins. to 7 ins. Fig. 48 is a detailed drawing giving principal dimensions of this pump.

Calculation of Power to Drive Pump and the Over-all Efficiency.

Actual quantity dealt with by runner is 2·94 cub. ft. per second. Work done per pound of fluid passing through runner is $H_t = 85\cdot7$ ft. lb.

$$N = \frac{H_t Q_t G}{550} = \frac{85\cdot7 \times 2\cdot94 \times 62\cdot4}{550} = 28\cdot5.$$

Power absorbed in overcoming friction of runner is

$$N_r = 0\cdot0000087 u_2^{3} r^2.$$
$$= 0\cdot0000087 \times 70\cdot5^3 \times 0\cdot562^2 = 0\cdot962.$$

Power to overcome friction of bearings N_b is

$$N_b = 0\cdot0000554 d^3 n.$$
$$= 0\cdot0000554 \times 5\cdot06 \times 1200 = 0\cdot337.$$

Total power to drive pump $N = N_t + N_r + N_b$
$$= 28\cdot5 + 0\cdot962 + 0\cdot337$$
$$= \text{say } 29\cdot8.$$

Short-circuit loss must now be estimated.

The difference of head between outside of runner and inlet is $0\cdot7\dfrac{u_2^{2}}{2g} = 54$ ft.

Taking velocity of flow through joint as v_1

$$v_1 = \tfrac{1}{2}\sqrt{2gh'} \text{ where } h' = 54 \text{ ft.}$$
$$= 4\sqrt{54} = 29\cdot4 \text{ ft. per second.}$$

If clearance is $\frac{1}{64}$ in.,

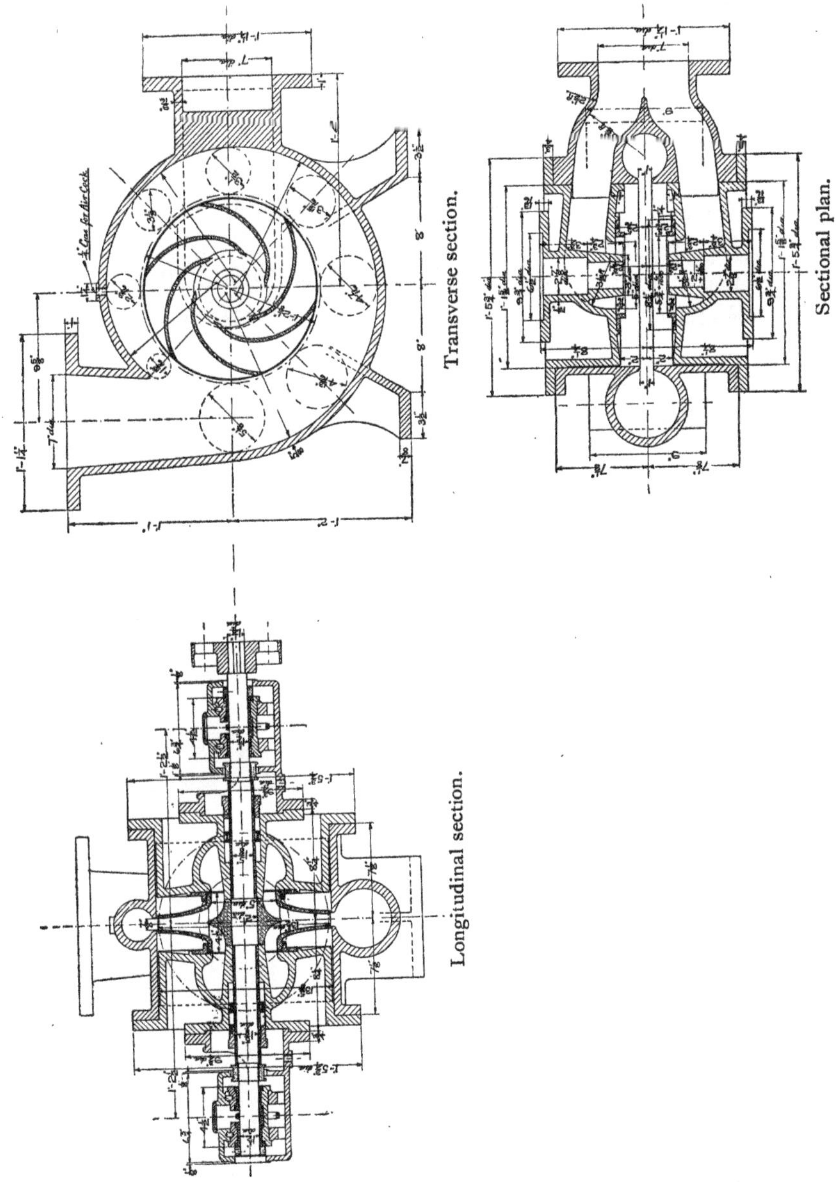

Transverse section.

Sectional plan.

Longitudinal section.

Fig. 48.—Design for a single-stage pump.

Area across which leakage takes place is

$$\frac{\pi 5\cdot75 \times 1}{144 \times 64} = 0\cdot00196 \text{ sq. ft.}$$

Since there are inlets to runner, short-circuit loss will be

$$2 \times 0\cdot00196 \times 29\cdot4 = 0\cdot115 \text{ ft.}^3 \text{ per second.}$$

Therefore total quantity discharged by pump is $2\cdot94 - 0\cdot115$ $= 2\cdot825$.

The quantity allowed was $2\cdot94 - 2\cdot67 = 0\cdot27$ cub. ft.

Water to seal glands would amount to $0\cdot035$ cub. ft. per second.

Total water discharged is $2\cdot79$ cub. ft. per second or 1045 gallons per minute.

$$N_w = \frac{H_m QG}{550} = \frac{60 \times 2\cdot79 \times 62\cdot4}{550} = 19.$$

Over-all efficiency $= \eta = \dfrac{N_w}{N + N_r + N_b} = \dfrac{19}{29\cdot8} = 64$ per cent.

CHAPTER X.

CALCULATION AND DESIGN OF MULTI-STAGE PUMPS.

PUMP for 400 gallons per minute, 1000 ft. head, to be driven by an alternating current motor running at 1450 revolutions or 2900 revolutions per minute :—

First Assumption.—A six-stage pump running at a speed of 1450 revolutions per minute with a hydraulic balance plate.

$$Q = \frac{400}{60 \times 6\cdot24} = 1\cdot07 \text{ cub. ft. per second.}$$

Allowing 20 per cent. in this case for short-circuit losses and leakage by the balance plate or piston,

Then $Q_t = 1\cdot07 + 0\cdot214 = 1\cdot284$.

Head per stage $h = \dfrac{H}{6} = \dfrac{1000}{6} = 166\cdot66$ ft.

Angular velocity $\omega = 152$.

Assuming $\beta_2 = 30°$, $x = 0\cdot08$, since the quantity is relatively small compared with the head, and $\eta_h = 0\cdot8$,

$\sin \beta_2 = 0\cdot5$, $\cos \beta_2 = 0\cdot866$, $\cotan \beta_2 = 1\cdot732$.

$h_t = \dfrac{166\cdot6}{0\cdot8} = \text{say } 209$ ft.

$$u_2 = \sqrt{\frac{gh_t}{1 - \cot \beta_2 x}} = \sqrt{\frac{32\cdot2 \times 209}{1 - 1\cdot732 \times 0\cdot08}} = \sqrt{\frac{32\cdot2 \times 209}{0\cdot8614}}$$

$u_2 = \sqrt{7800} = 88\cdot3$ ft. per second.

$w_2 \sin \beta_2 = 0\cdot08 \times 88\cdot3 = 7\cdot064$ ft. per second.

$$w_2 = \frac{7\cdot064}{0\cdot5} = 14\cdot128 \text{ ft. per second.}$$

$r_2 = \dfrac{88\cdot3}{152} = 0\cdot581$ ft., $D_2 = 13\cdot95$, say 14 ins.

(87)

Making $r_2 = 0.583$ ft.

Supposing $\dfrac{p_2}{p_2 + t_2} = 0.9,$

$$b_2 = \frac{Q_t}{2\pi r_2 x u_2 \times 0.9} = \frac{1.284}{2\pi \times 0.583 \times 0.08 \times 88.3 \times 0.9} = 0.0553 \text{ ft.}$$

0.6636 ins. wide. $\qquad \dfrac{b_2}{r_2} = \dfrac{0.6636}{7} = 0.095.$

Since $\dfrac{b_2}{r_2}$ is less than 0.1, the value of $\eta_h = 0.8$ will not be attained, therefore the above calculations must be gone through again, using a lower value of η_h, say 0.77, or if a very high efficiency is desired, a larger number of stages must be used, say 8 instead of 6, or the speed increased to 2900 revolutions per minute.

Second Approximation.—Assuming 6 stages and $\eta_h = 0.77$, $x = 0.08$, $\beta_2 = 30°$.

$h = 166.6$ ft.

$$h_t = \frac{166.6}{0.77} = \text{say } 217 \text{ ft.}$$

$$u_2 = \sqrt{\frac{32.2 \times 217}{1 - 1.732 \times 0.08}} = \sqrt{\frac{32.2 \times 217}{0.8614}} = \sqrt{8100}.$$

$u_2 = 90$ ft. per second, $w_2 \sin \beta_2 = 7.2$ ft. per second.

$$r_2 = \frac{90}{152} = 0.593 \text{ ft.}, \; D_2 = 14\tfrac{1}{4} \text{ ins. diameter.}$$

$$b_2 = \frac{1.284}{2\pi \times 0.593 \times 0.08 \times 90 \times 0.9} = 0.0532 \text{ ft. or } 0.6384 \text{ in.,}$$

say $\tfrac{5}{8}$ in.

$$\frac{b_2}{r_2} = \frac{0.625}{7.125} = 0.0914.$$

Diameter of shaft in coupling $= 1\tfrac{5}{8}$ in.

Diameter of shaft in bearing $= 1\tfrac{7}{8}$ in.

Disc friction $N_r = 9.48$ horse-power.

Bearing friction $N_b = 0.53$.

$N_t =$ horse-power expended on water passing through runners.

$$N_t = \frac{6 \times 217 \times 62.4 \times 1.284}{550} = 189.5.$$

Power to drive pump $N = N_t + N_r + N_b$
$$= 189\cdot5 + 9\cdot48 + 0\cdot53$$
$$= \underline{199\cdot5.}$$

Water horse power $N_w = 121.$

$$\text{Efficiency} = \frac{N_w}{N} = \frac{121}{199\cdot5} = 0\cdot607 \text{ or } 60\cdot7 \text{ per cent.}$$

This is the highest efficiency attainable with the above proportions. If a higher efficiency is required without increasing the speed, the number of stages must be increased.

Third Approximation.—Assuming 8 stages, running at 1450 revolutions per minute, and a provisional value of

$\eta_h = 0\cdot81.$

$B_2 = 30°,\ x = 0\cdot08,\ Q_t = 1\cdot284.$

$$h = \frac{1000}{8} = 125 \text{ ft.}$$

$$h_t = \frac{125}{0\cdot81} = \text{say } 154 \text{ ft.}$$

$$u_2 = \sqrt{\frac{gh_t}{1 - x \cot B_2}} = \sqrt{\frac{32\cdot2 \times 154}{0\cdot8614}} = 75\cdot8 \text{ ft. per second.}$$

$$r_2 = \frac{75\cdot8}{152} = 0\cdot5,\ D_2 = 12 \text{ ins.}$$

$w_2 \sin \beta_2 = 6\cdot07$ ft. per second.

$$b_2 = \frac{1\cdot284}{2\pi \times 0\cdot5 \times 6\cdot07 \times 0\cdot9} = 0\cdot075 \text{ ft. or } 0\cdot9 \text{ in. or } \frac{25}{32} \text{ in.}$$

$$\frac{b_2}{r_2} = \frac{0\cdot9}{6 \text{ ins.}} = 0\cdot15.$$

Entrance velocity $= 9$ ft. per second.

$$\text{Area entrance} = \frac{1\cdot284}{9} = 0\cdot143 \text{ sq. ft. or } 20\cdot6 \text{ sq. ins.}$$

Diameter of eye $= 6\frac{3}{8}$ ins.

Diameter of inlet edge of vanes $= 6\frac{1}{2}$ ins.

$$u_1 = 41\cdot2 \text{ ft. per second.}$$

$$\tan \beta_1 = \frac{9}{41\cdot2} = 0\cdot219$$

$$\beta_1 = 12° \ 30'.$$

End thrust to right due to change of direction of inlet

$$\text{velocity} = 8\frac{\pi}{4}(d_2{}^2 - d_4{}^2)\frac{v^2}{g} = 8 \times (31\cdot9 - 11)\frac{81}{32\cdot2} = 8 \times 20\cdot9 \times \frac{81}{32\cdot2}$$

$$= 420 \text{ lb.}$$

$$\text{Thrust to left} = 8(38\cdot485 - 8\cdot295)\frac{69\cdot5}{2\cdot3} = 8 \times 30\cdot19 \times \frac{69\cdot5}{2\cdot3}$$

$$= 7290 \text{ lb.}$$

$$T = 7290 - 420 = 6870 \text{ lb.}$$

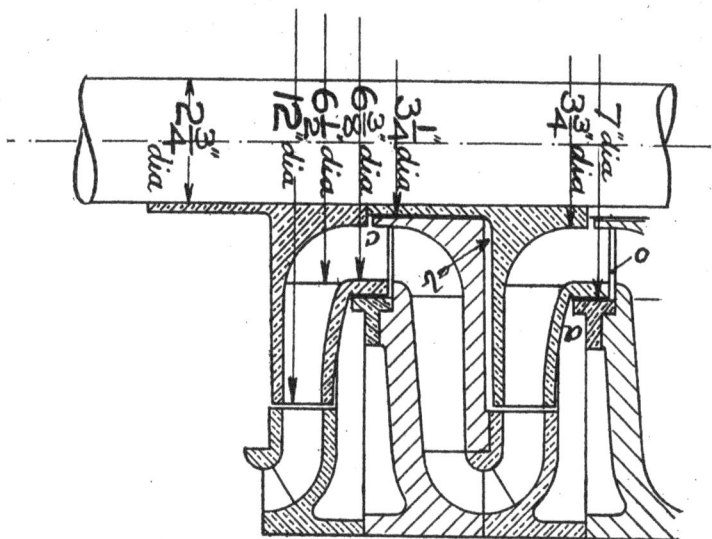

FIG. 48A.—Section of runner and guide, multi-stage pump.

$$\text{Pressure from last runner} = \frac{944\cdot5}{2\cdot3},$$

$$\text{area piston} = \frac{6870 \times 23}{944\cdot5} = 16\cdot7$$

$$\frac{\pi}{4}d_p{}^2 = 16\cdot7 + 8\cdot295 = 24\cdot995$$

$$d_p = 5\tfrac{5}{8}$$

outer diameter $= 6\frac{3}{4}$ ins. giving a bearing face of $\frac{9}{16}$ in.

Leakage from a to o.

Difference of head producing leakage 69·5 ft.

Clearance at joint $= 0.004$.

Area through which leakage takes place

$$= \frac{\pi 7 \times 0.004}{144} = \frac{0.088}{144}.$$

Velocity $= \frac{1}{2}\sqrt{2g69.5} = 4\sqrt{69.5} = 33.3$.

Leakage $= \dfrac{0.088}{144} \times 33.3 = 0.02$ cub. ft. per second.

Leakage from c to b.

Area $= \dfrac{\pi 3.25 \times 0.004}{144} = \dfrac{0.041}{144}.$

Head producing flow $= 125 - 69.5 = 55.5$.

$v = \frac{1}{2}\sqrt{2g55.5} = 4\sqrt{55.5} = 29.8$.

Leakage $= \dfrac{0.041}{144} \times 29.8 = 0.0085$ cub. ft. per second.

Leakage past balance piston or disc.

Head producing flow $= 944.5$ ft.

$$v = 0.6\sqrt{2g944.5} = 0.6 \times 8\sqrt{944.5} = 148 \text{ ft. per second.}$$

Area past which leakage occurs $= \dfrac{\pi 6.75 \times 0.004}{144} = \dfrac{0.085}{144}.$

Leakage $= \dfrac{0.085 \times 148}{144} = 0.0875$.

Total leakage and short circuit $= 0.0875 + 0.0085 + 0.02$
$= 0.098$, say 0.1 cub. ft. per second.

Water actually pumped $= 1.284 - 0.1$
$$= 1.184 \text{ cub. ft. per second.}$$

$N_r = 7.949$.

$N_b = 0.53$.

$N_t = 179.5$, i.e. $\dfrac{1.284 \times 8 \times 154 \times 62.4}{550}.$

$N = N_r + N_b + N_t = 188$.

$N_w = \dfrac{1.184 \times 1000 \times 62.4}{550} = 134$.

Efficiency $= \dfrac{N_w}{N} = \dfrac{134}{188} = 0.715$ or 71.5 per cent.

For an eight-stage pump dealing with 400 gallons per minute against 1000 ft. running at 1450 revolutions per minute the efficiency would be about 71·5 per cent. Therefore if a higher efficiency is desired, since the hydraulic efficiency cannot be greatly improved upon, it will be necessary to reduce the power expended in overcoming the friction of the runner sides, and to do this it will be necessary to run at a higher speed than 1450 revolutions.

Assuming the number of stages as 5, and the speed 2900 revolutions per minute, the leading dimensions of a pump to perform the above duty will be calculated

$$\omega = 2900 \times \frac{\pi}{30} = 304. \qquad h = \frac{1000}{5} = 200 \text{ ft.}$$

$$\beta_2 = 30°.$$

Assuming $\eta_h = 0·83$ and $x = 0·1$.

$$h_t = \frac{200}{0·83} = 241. \qquad Q_t = 1·284 \text{ cub. ft. per second.}$$

$$u_2 = \sqrt{\frac{32·2 \times 241}{0·8268}} = 93·8 \text{ ft. per second.}$$

$$r_2 = \frac{93·8}{304} = 0·3083 \text{ ft.}$$

$d_2 = $ say 7·4375 ins. or $7\frac{7}{16}$ ins. ; this gives $r_2 = 0·3099$ ft.

$w_2 \sin \beta_2 = xu_2 = 0·1 \times 93·8 = 9·38$ ft. per second.

$$b_2 = \frac{1·284}{2\pi \times 0·3099 \times 9·38 \times 0·9} = 0·078 \text{ ft. or say } \frac{15}{16} \text{ in.}$$

$$\frac{b_2}{r_2} = \frac{0·078}{0·3099} = 0·252.$$

This value for $\frac{b_2}{r_2}$ justifies the assumption of taking $\eta_h = 0·83$.

$w_1 \sin \beta_1 = 14$ ft. per second.

Entrance area $= \dfrac{1·284}{14} = 0·0918$ sq. ft. $= 12·85$ sq. ins.

Area of boss $= \dfrac{\pi}{4} 3·25^2 = 8·3$ sq. ins.

$$\frac{\pi}{4} D_e^2 = 12·85 + 8·3 = 21·15 \text{ sq. ins.}$$

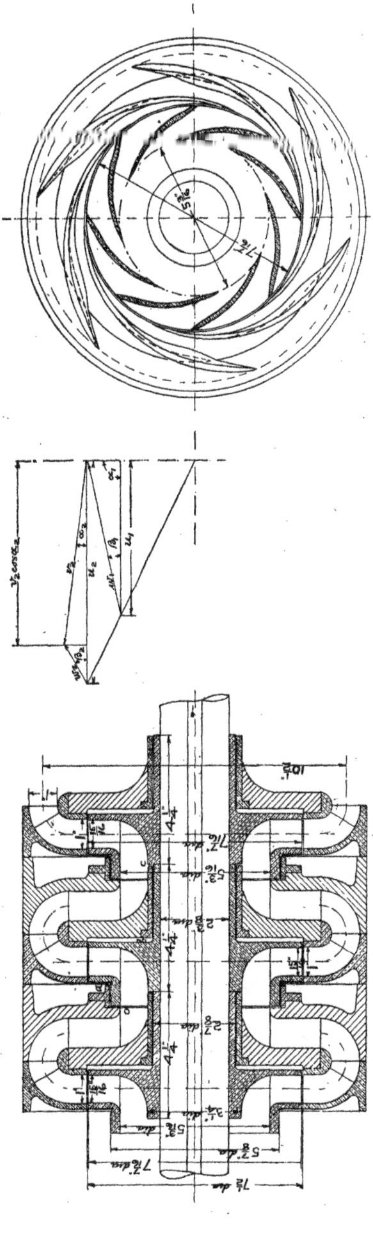

Section through runner and guide vanes.

Cross section.

FIG. 49.—Design for runner and guides for five-stage pump.

Diameter of entrance = say $5\frac{5}{16}$ ins.

$$u_1 = 93\cdot8 \times \frac{5\cdot1875}{7\cdot4375} = 65\cdot5 \text{ ft. per second.}$$

$$\tan \beta_1 = \frac{w_1 \sin \beta_1}{u_1} = \frac{14}{65\cdot5} = 0\cdot214$$

$$\beta_1 = 12° \ 5'.$$

The method for drawing the runner vanes is the same as already explained for the single-stage pump, but in this case there is only one inlet. Fig. 49 shows details of runners and guides, while Fig. 50 is a sectional drawing of the complete pump with the principal dimensions.

Calculation of Balance Disc.

Forces acting to the right on rotor are equal to

$$5\frac{\pi}{4}(d_2{}^2 - d_4{}^2)\frac{v^2}{g} = 5\frac{\pi}{4}(5\tfrac{3}{16}{}^2 - 3\tfrac{3}{8}{}^2)\frac{14}{32} = 373 \text{ lb.}$$

Forces acting to the left on rotor are

$$5\frac{\pi}{4}(d_3{}^2 - d_4{}^2)\frac{110}{2\cdot3} \text{ ft.} = (27\cdot1 - 6\cdot49)\frac{5 \times 110}{2\cdot3} = 4920 \text{ lb.}$$

Unbalanced force to left = 4920 − 373
$$= 4547 \text{ lb.}$$

Pressure from last runner is $\dfrac{910}{2\cdot3}$ ft. or 396 lb. per sq. in. and is the value to be taken for px.

Area of balance plate or disc $= \dfrac{4547}{396} = 11\cdot5$ sq. ins.

Diameter of balance is dp.

$$\frac{\pi}{4}d_p{}^2 = \frac{\pi}{4}d_4{}^2 + 11\cdot5 = 6\cdot49 + 11\cdot5 = 17\cdot99 \text{ sq. ins.}$$

$$d_p = \text{say } 4\tfrac{3}{4} \text{ ins.}$$

Making outer diameter equal to 6 ins., the width of the working face will be $\frac{5}{8}$ in.

Head producing flow past disc is 910 ft.

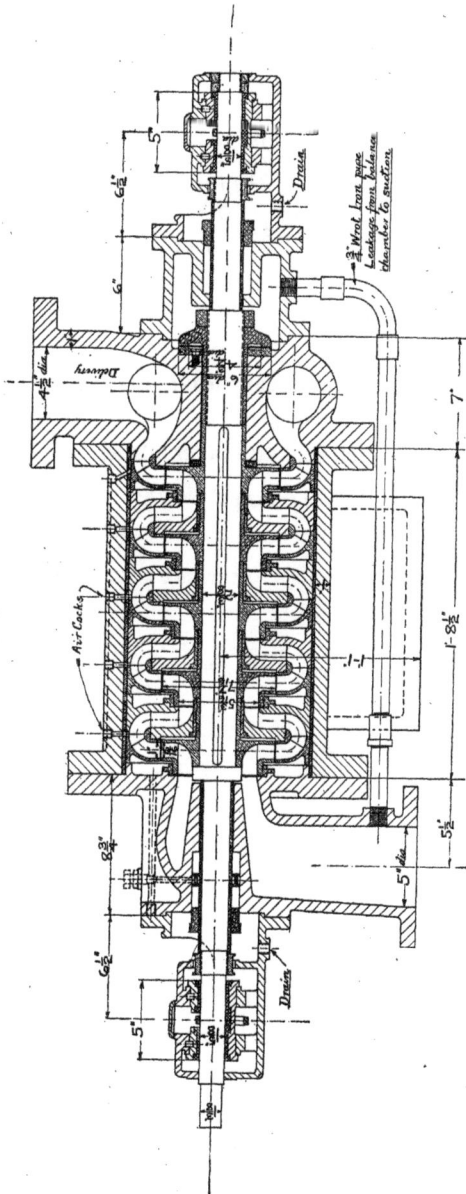

FIG. 50.—Design for five-stage pump.

Velocity of flow is $0.6\sqrt{2g910} = 145$ ft. per second.

0.6 being a coefficient to allow for friction.

If the clearance between face of disc and its pad is 0.004 in.,

$$\text{area of outlet} = \frac{\pi 6 \times 0.004}{144} = \frac{0.0755}{144} \text{ sq. ft.}$$

Leakage past balance disc $= \dfrac{0.0755}{144} \times 145 = 0.076$ cub. ft. per second.

Short-circuit losses—

Leakage from a to o.

Difference of head producing leakage $= 110$ ft.

Clearance at joint $= 0.004$ in.

Area through which leakage takes place.

$$= \frac{\pi 5.875 \times 0.004}{144} = \frac{0.0737}{144} \text{ sq. ft.}$$

Velocity of fluid through clearance $= \frac{1}{2}\sqrt{2g110} = 42$ ft. per second.

Short-circuit loss from a to $o = \dfrac{0.0737 \times 42}{144} = 0.0215$ cub. ft. per second.

Leakage from c to b.

Clearance $= 0.004$.

Area through which leakage takes place $= \dfrac{\pi 2.875 \times 0.004}{144}$

$$= \frac{0.036}{144} \text{ sq. ft.}$$

Head producing leakage is the gain of pressure head in diffuser and is equal to $200 - 110 = 90$ ft.

Velocity of flow $= \frac{1}{2}\sqrt{2g90} =$ say 38 ft. per second.

Short-circuit loss is $\dfrac{0.036 \times 38}{144} = 0.0095$.

Total short-circuit and leakage loss $= 0.076 + 0.0215 + 0.0095$
$$= 0.107 \text{ cub. ft. per second.}$$

Say 0.11 cub. ft. per second to allow for gland seal.

Water actually pumped $= 1\cdot284 - 0\cdot11$
$$= 1\cdot174 \text{ cub. ft. per second.}$$

$$N_t = \frac{241 \times 1\cdot284 \times 62\cdot4 \times 5}{550} = 176.$$

$N_r = 3\cdot449.$

If shaft is made $1\frac{5}{8}$ ins. diameter in the bearings and $1\frac{3}{8}$ ins. diameter in coupling,

$N_b = 0\cdot69.$

Power to drive pump $N = 176 + 3\cdot449 + 0\cdot69 = 180\cdot14.$

$$N_w = \frac{1\cdot174 \times 1000 \times 62\cdot4}{550} = 133.$$

Efficiency $= \dfrac{N_w}{N} = \dfrac{133}{180} = 0\cdot74$ or 74 per cent.

This increase in efficiency over that of the previous example is principally due to the decrease in the friction of the runner sides.

CHAPTER XI.

THE following illustrations show the pumps designed and built by the American Well Works of Aurora, Illinois, U.S.A.

Fig. 51 shows one of their volute type pumps classified as Type DS Single-Stage "American" Double Suction, which is one of their latest types, and is used for heads up to 125 ft.

The fluid pumped enters the runner from two sides and is discharged into a well-designed volute chamber.

The shaft is of steel and is fitted with renewable brass liners so that the steel does not come in contact with the fluid pumped.

The shaft is supported in ring oiled bearings outside the pump, which are designed so that the fluid pumped does not enter them.

A special feature of these pumps is the renewable labyrinth packing rings, these rings being so designed as to reduce the short-circuit losses from the discharge chamber into the suction chamber to the least possible amount, thus enabling this type of pump to maintain its efficiency. These rings have already been shown in Fig. 35 for the case of single inlet runners. The volute casing is in halves, being split horizontally, and removal of the top half of casing and bearing caps enables the whole rotor complete, including shaft and runner, to be removed for examination or renewal.

Fig. 52 shows a characteristic curve of a pump of this type, Fig. 51 having 15 ins. diameter suction inlet and discharge orifice.

The efficiency of the pump is 80 per cent. when delivering

(98)

5000 U.S. gallons per minute (4150 Imperial gallons per minute) against a total head of 18 ft.

The rise of head above the normal 18 ft. when a sluice valve on the delivery is closed, no delivery then taking place, is 22·8 ft. or 1·27 times the normal head.

It will be seen that the power curve rises to maximum at 7200 U.S. gallons per minute (6000 Imperial gallons per minute) and then tends to decrease for any further increase in the quantity, the pump itself being thus self-regulating as regards power.

This latter is considered by some an important point with electrically driven pumps, although the switch gears of motors have an arrangement for cutting off the current and stopping the motor; this would only happen by the head being reduced, due to a breakage in

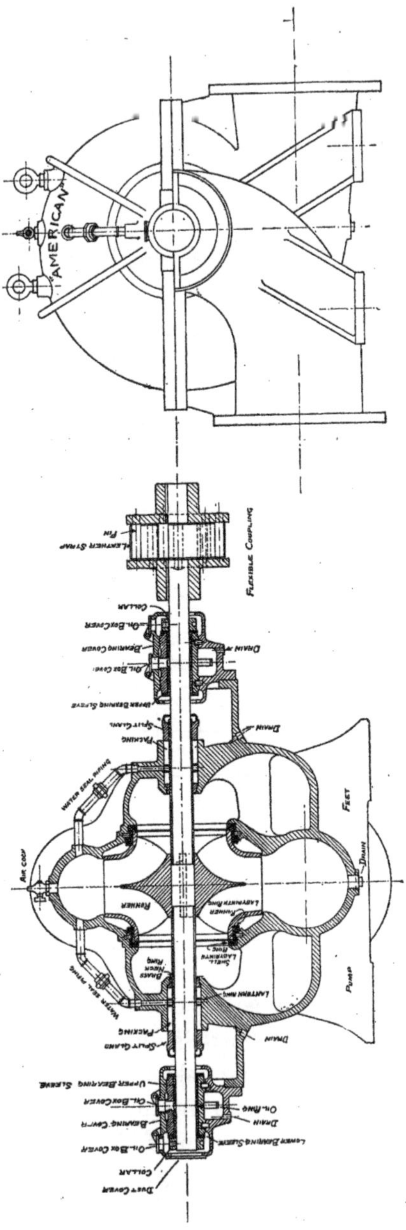

Fig. 51.—Single-stage pump.

the delivery pipe line near the pump, which would be a very
unusual incident

Fig. 46, chapter VIII., shows a two-stage pump classified as
Type K.M.B. Two-Stage American Centrifugal Pump, whilst
Fig. 53 represents the characteristic curves of test of a two-stage
5-in. pump of this type. The runners in this pump are of the

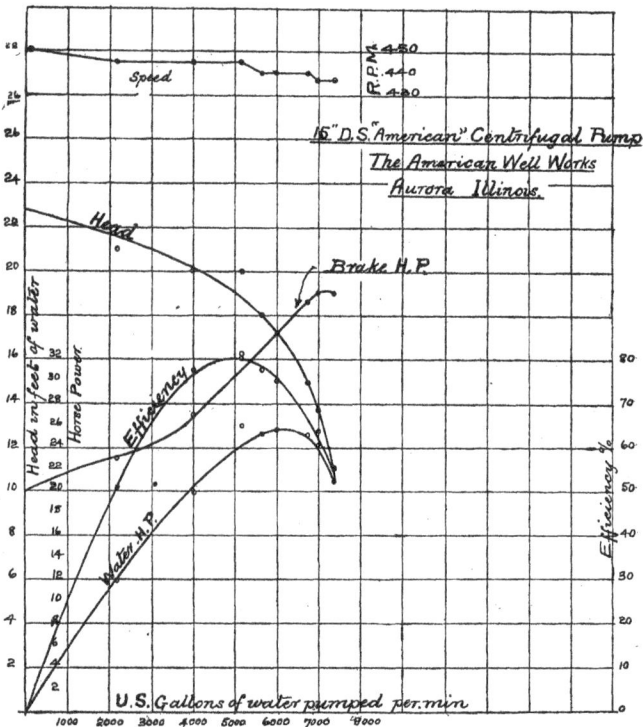

FIG. 52.—Characteristic curves for single-stage pump.

single inlet type, fitted with labyrinth renewable packing rings
which reduce short-circuit losses to a minimum, the unbalanced
axial thrust being taken up by a plate or piston, the leakage
water returning to the suction inlet by a byepass.

For a 5-in. pump of this type Fig. 53 shows that when
dealing with 700 gallons U.S. per minute against a head of

136 ft. when running at a speed of 1300 revolutions per minute the efficiency is 71 per cent.

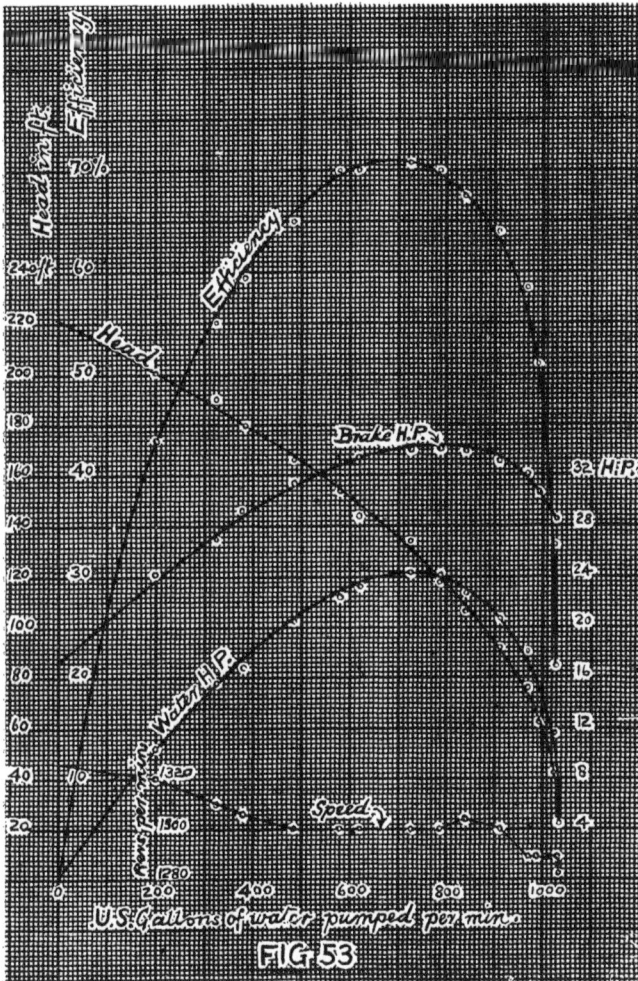

FIG. 53.—Characteristic curves, two-stage pump.

A very interesting product of this firm is their application of the centrifugal pump as a bore well pump, known as The

"American" Combination Deep Well Turbine and Booster Centrifugal.

Figs. 54 and 55 show this combination driven by a vertical spindle electric motor, and installed by this firm for the City of Rockford, Illinois, U.S.A.

The combination furnished consists of a 17-in. six-stage combination deep well and pressure pump, being a four-stage turbine type of pump located 108 ft. down in the bore well, with a two-stage centrifugal type of pump at the top of well, as shown in Figs. 54 and 55, the guaranteed over-all operating efficiency. to be 59½ per cent.

Fig. 54 is an exterior view of a similar pump to the one installed at Rockford, whilst Fig. 55 shows sectional and outside views of the Rockford pump. It will be noticed that water from the bottom pump flows up the annular space between two pipes, thus avoiding the friction losses which would occur if such a long shaft revolved in water.

On the several tests run after this plant was installed the pump proved to have an efficiency of 65·5 per cent. when delivering 1380 U.S. gallons per minute against a total head of 237·57 ft.

FIG. 54.—Bore well pump.

The following is a facsimile of the last test made :—

Capacity, 1380 U.S. gallons per minute.

Discharge pressure 61·5 lb. per sq. in., equal to 142 ft.

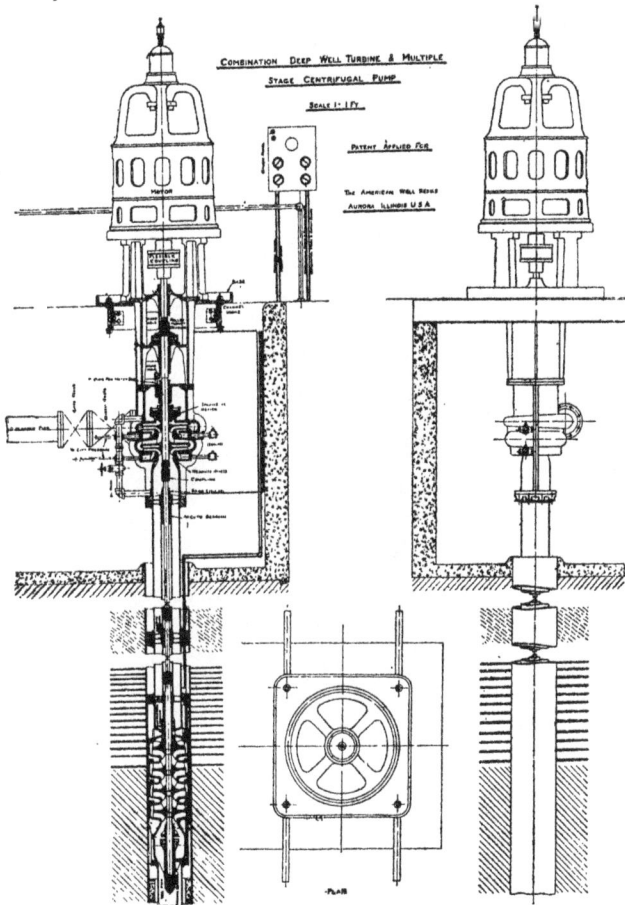

FIG. 55.—Section of bore well pump.

Static head to centre of pressure gauge, 95·57 ft.

Total head of, 237·57 ft.

$$\text{Water horse-power} = \frac{1380 \times s \times 237\cdot57}{33{,}000} = 82\cdot1.$$

Meter reading ahead of transformers, 116 kilowatts.

90 per cent. efficiency of transformers, $116 \times 0\cdot9 = 104\cdot4$ K.W.

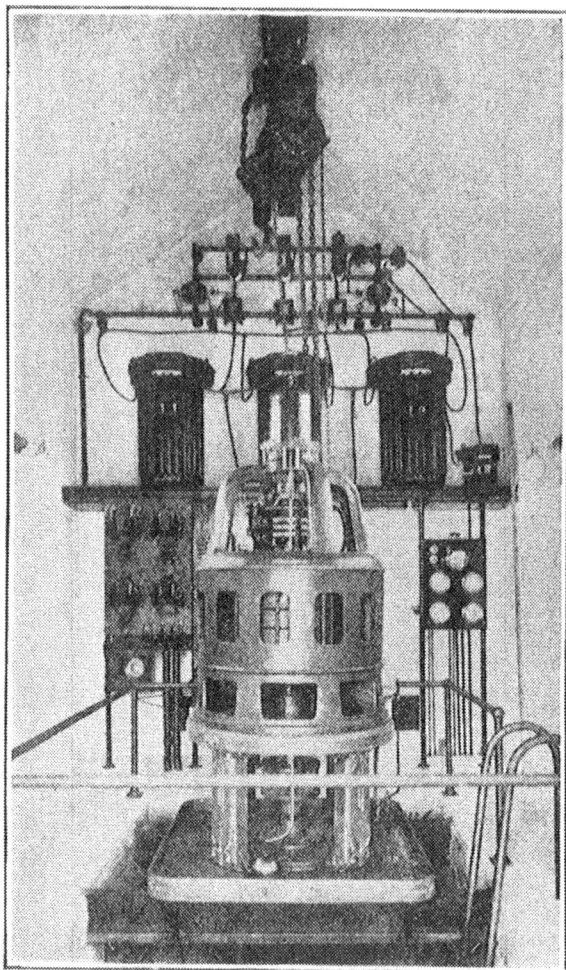

FIG. 56.—Electric motor driving bore well pump.

90 per cent. efficiency of motor, $104\cdot4 \times 0\cdot9 = 93\cdot39$ K.W. or 125 H.P. delivered to pump shaft.

Efficiency of pump $= \dfrac{N_w}{N_p} = \dfrac{82 \cdot 1}{125} = 0 \cdot 655$ or $65 \cdot 5$ per cent.

Fig. 56 is a view of the interior of the pump house, showing detail of wiring and travelling crane.

FIG. 57.—View of delivery piping at top of bore well pump.

Fig. 57 is an interior view of pump house, showing vertical centrifugal pump and piping at top of shaft in the pit immediately below motor.

Fig. 58 gives the characteristic curves, showing relation

between capacity, head, theoretical horse-power, brake horse-power, and efficiency of pump.

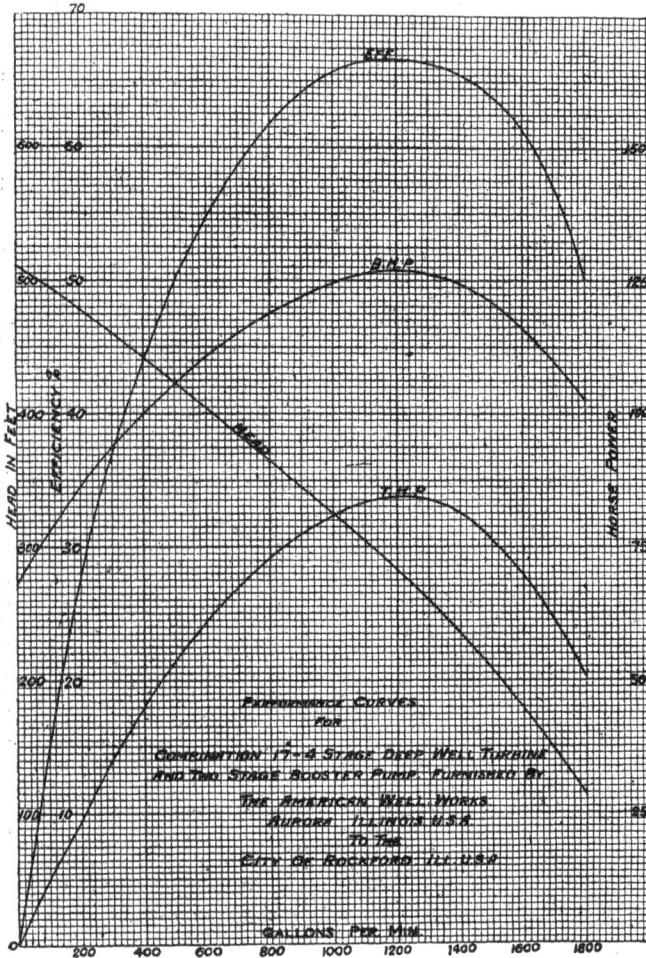

FIG. 58.—Characteristic curves showing relation between capacity, head, theoretical, horse power, brake horse power and efficiency on pump installed in well No. 7, at Rockford, Illinois.

Fig. 59 is an illustration of a single-stage American centrifugal pump direct coupled to a Kerr steam turbine, whilst Fig.

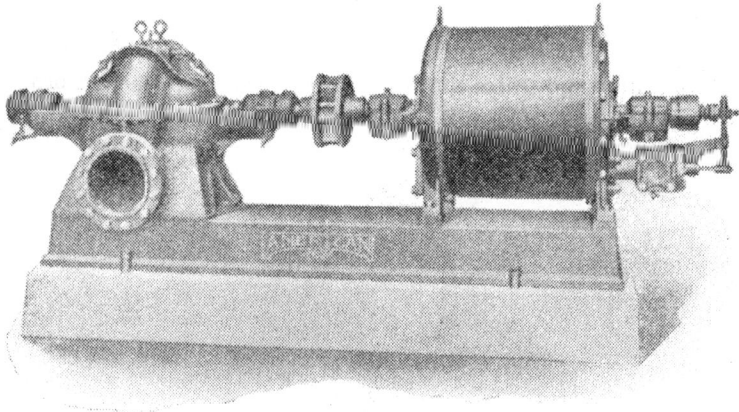

FIG. 59.—Single-stage pump driven by steam turbine.

60 shows a high pressure single-stage pump coupled to a Terry steam turbine supplied to the Omaha Packing Co., Chicago.

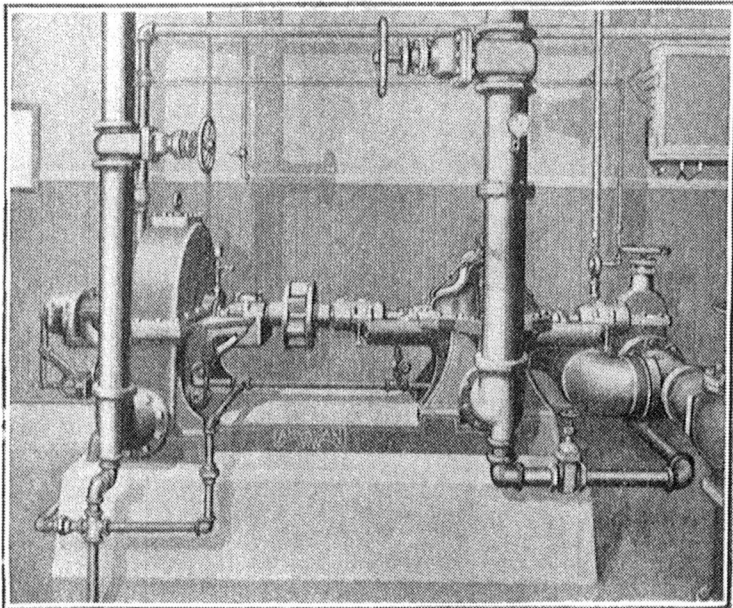

FIG. 60.—Single-stage pump driven by steam turbine.

The Sulzer Pumps.

Messrs. Sulzer Bros. of Winterthur, Switzerland, were the first firm to recognize that guide vanes were necessary in order to obtain good efficiency in centrifugal pumps.

FIG. 61.—Electrically driven centrifugal.

FIG. 62.—View of multi-stage pump dismantled.

They were also the first to satisfactorily overcome the evils due to axial thrust in multi-stage pumps.

This was accomplished in their early pumps by arranging the runners back to back in pairs.

The fluid from exit of the guide passages in one stage, to the inlet of the runner in the next stage, is carried through specially constructed passages, passing through the diffuser between its guide passages.

As this arrangement necessitates rather complicated castings, they now adopt a design somewhat similar to Fig. 45, and balance the end thrust by means of a balance plate, of which they are the originators.

Fig. 61 illustrates a horizontal multi-stage pump by this firm, direct coupled to an electric motor, and Fig. 62 the pump dismantled.

Fig. 63 presents characteristic curves showing the performance of one of their five-stage pumps with constant delivery sluice valve opening, showing variation of head, quantity, power, and speed, the efficiency remaining constant at 77 per cent.

The characteristic curves in Fig. 64 for the same pump show the variation in head, quantity, power and efficiency when the speed is constant, the variation in head and quantity being obtained by throttling the discharge by means of the sluice valve on the delivery.

The maximum efficiency is 77 per cent., the total head being 1140 ft., quantity 1185 gallons per minute, the power to drive the pump being 542 at 1500 revolutions per minute.

It will be noticed that this pump could be used over a considerable range without a great drop in the efficiency.

For instance, when pumping 1000 gallons per minute against 1190 ft., at 1500 revolutions per minute, the efficiency would be about 75 per cent., or a drop of about 2 per cent.

Also, when 1400 gallons per minute against 1040 ft. at 1500 revolutions, the efficiency would be about 73 per cent. This shows that the pump can be used over a considerable range with very little loss in efficiency.

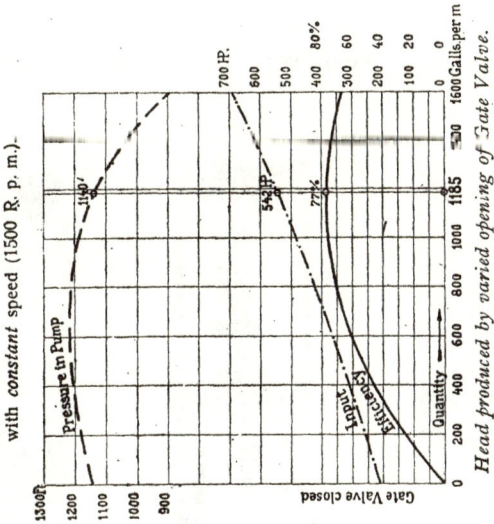

Head produced by varied opening of Gate Valve.
FIG. 64.

Working conditions of a H.C.P. 5 stage No. IV B e

with constant speed (1500 R. p. m.).

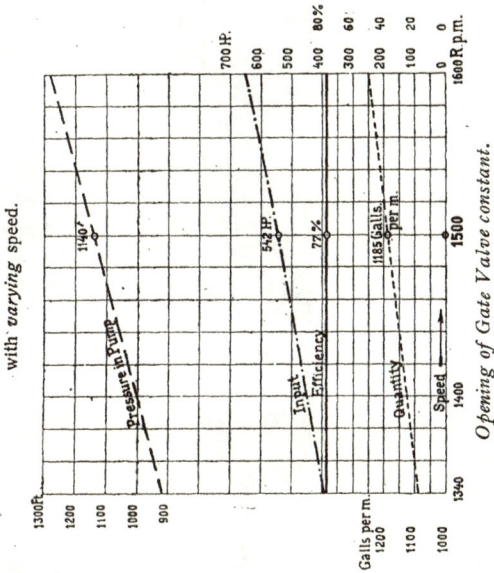

Opening of Gate Valve constant.
FIG. 63.

with varying speed.

Characteristic curves.

Vertical spindle centrifugal pumps coupled direct to an
electric motor, both being mounted together in a sliding frame,

and suspended so as to be movable in a vertical direction, are used as sinking pumps.

These pumps take up very little space in comparison with

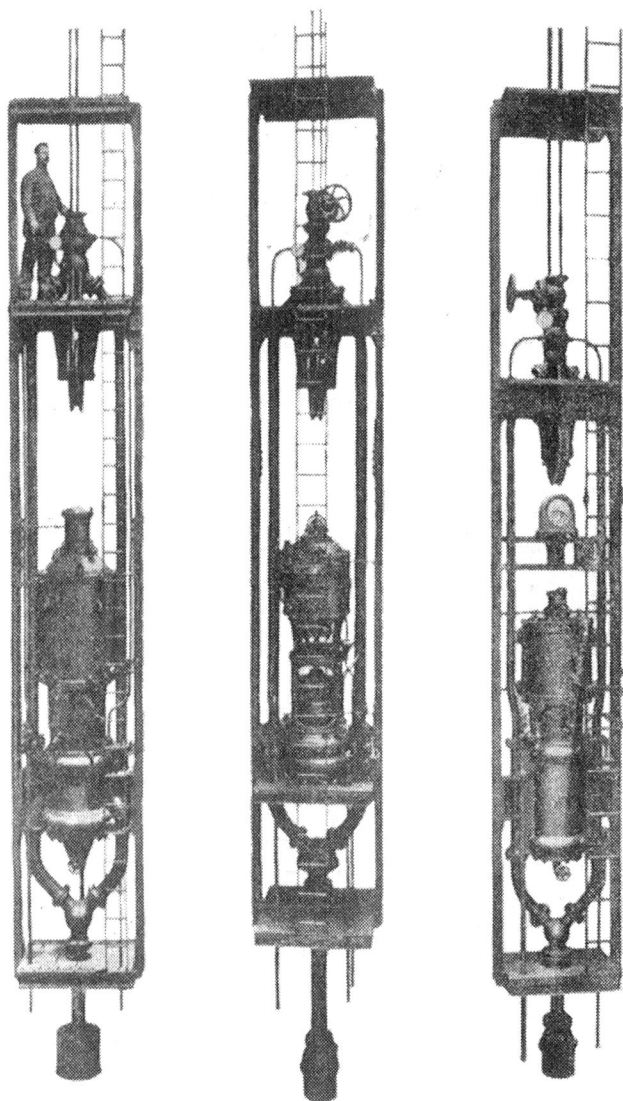

FIG. 65.—Centrifugal sinking pumps.

their output, and are specially constructed for sinking new mine shafts, as well as unwatering existing ones.

FIG. 66.—Centrifugal driven by steam turbine.

Upon completion of these operations they may be used as stationary pumps.

Fig. 65 represents a sinking pump made by this firm.

Messrs. Sulzer Bros. have constructed a considerable number of high lift centrifugal or turbine pumps for the water supply of towns. The pumps for this purpose are driven by electric motors, steam engines, steam turbines and internal combustion engines.

The pump in Fig. 66, direct coupled to a steam turbine, is one of four supplied to the St. Petersburg Water Works, each pump dealing with 9460 gallons of water per minute against a head of 164 ft.

An electrically driven plant for water supply is shown in Fig. 67.

This plant, for the city of Lyons, consists of 3 pumps, each 2800 gallons per minute, 165 ft. head; 1 pump, 1100 gallons per minute, 330 ft. head; 2 pumps, each 800 gallons per minute, 550 ft. head; and 2 pumps, each 860 gallons per minute, 445 ft. head.

Fig. 68 is a pump driven by belt from a Diesel engine, the pump delivering 1300 gallons per minute against 1200 ft. head.

Wherever high pressure water is required for working hydraulic plants (lifts, presses, etc.), and hydraulic transmission of power, the turbine pump can be used advantageously, especially for large plants.

Although the requirements of high pressure water fluctuates greatly in such installations, these fluctuations may be considerable without greatly affecting the efficiency of the pump.

The pump can be designed so that the quantity may vary considerably, with very small variation in the delivery pressure, the pump thus accommodating itself automatically to the quantity of water required, although it is running at a constant speed.

Fig. 69 shows a pump for the hydraulic main supply of the port of Amsterdam for working cranes in the harbour, and delivers 220 gallons per minute at 740 lb. per sq. in., which is equivalent to a head of 1700 ft.

FIG. 67.—Centrifugal pumping plant.

The centrifugal or turbine pump in fig. 70, in conjunction with the plunger pumps, works a loaded accumulator.

The centrifugal pump in this case is doing alone the same amount of work as all the plunger pumps working together,

FIG. 68,—Centrifugal pump driven by Diesel engine.

and is a striking example of the small space required by the turbine pump.

FIG. 69.—Centrifugal pump with electric motor.

Other applications for these pumps are boiler feeding, fire extinguishing, supply of fountains, irrigation, and dock pumping.

FIG. 70.—High pressure centrifugal working in conjunction with reciprocating pumps for working lifts and presses, etc.

Pumps by Messrs. Hayward, Tyler & Co.

Messrs. Hayward, Tyler & Co. Ltd. of 99 Queen Victoria Street, London, have for some years built high lift centrifugal or turbo pumps known as the H.T. turbo-pump patented by Messrs. Gelpke & Kugel.

These pumps are of the mixed flow type, the flow through the runners and guides being first axial, then radial, and finally axial at discharge.

Fig. 71 is a section of a five-stage pump; the liquid enters the runner B in an axial direction and after flowing through the runner is discharged, also in an axial direction, into the stationary guide wheels C of same diameter as the runners B.

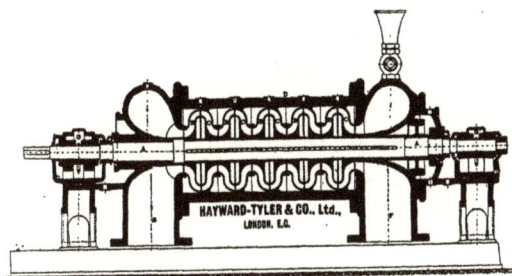

Fig. 71.—Section of H.T. turbo-pump.

The vanes in the guide wheels C are designed in such a manner that passages formed by two consecutive vanes gradually increase in area up to the entrance of the following runner.

The absolute velocity of the liquid is therefore gradually decreased, the kinetic energy of the fluid being transformed into pressure energy.

The guide vanes are so designed that this increase in area takes place uniformly in order to reduce eddy losses to a minimum.

The axial thrust is balanced by hydraulic pressure on a piston.

The advantage of this design of pump is that, owing to the form of the runners and guides, the diameter of the pump casing is considerably less than that of the purely radial type of the same output.

Also, owing to the entrance and discharge from the passages being axial, the shaft can have a considerable axial play, the runners always discharging directly into the guides.

Again, as entrance and discharge are in an axial direction the end thrust due to momentum at inlet and outlet is equal and opposite, therefore the difference of the end thrust due to pressure, only has to be balanced.

Due to these two latter conditions, no thrust block of the collar type is required or furnished with these pumps.

Also the liquid is under the control of the vanes during its entire passage through the pump, thus enabling the theoretical velocity diagrams to be realised.

Fig. 72 gives the front and back views of the runner and guide wheel, which are constructed of bronze, the vanes and passages being made perfectly smooth.

A good example of this design is illustrated in Fig. 73, which is a section of a six-stage pump arranged in a tube well.

Owing to the small external diameter of this type compared with the purely radial type, it was possible to install this pump, which deals with 350 gallons per minute, against a total head of 170 ft. when running at 1220 revolutions per minute in a tube well of 15 ins. internal diameter.

This pump was supplied to the Great Grimsby waterworks.

It is placed about 40 ft. down the tube well, the suction pipe descending 20 ft. below the pump.

The delivery from the pump passes up to the surface in the annular space between two pipes, as seen in Fig. 73, in lengths of about 10 ft. Each of these lengths of annular rising main is complete with its length of shaft with half couplings and bearings.

Owing to the discharge passing up the annular space, the shaft revolves in a perfectly dry space.

At the surface the annular space connects to a chamber, to which the delivery pipe to a water tower is connected.

This pump is driven by a horizontal gas engine through double helical machine-cut bevel wheels in one reduction, the

FIG. 72.

GUIDE

RUNNER

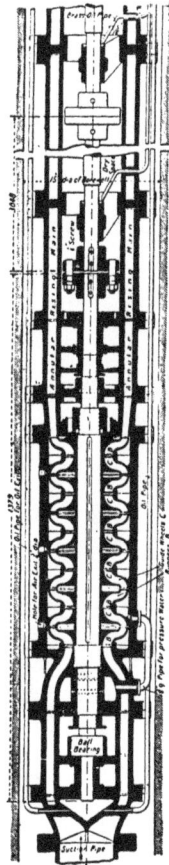

FIG. 73.—Section of bore well pump.

pinion being keyed to the top of the vertical pump shaft, with a ball bearing at the surface to carry the weight of the shaft only.

Fig. 74 is a three-stage pump direct coupled to an electric motor, and deals with 500 gallons of water per minute against

FIG. 74.—Electrically driven H.T. turbo-pump.

a vertical head of 147 ft. when running at 800 revolutions per minute.

The pump shown in Fig. 75 is a single-stage pump supplied to the Luton Corporation.

This pump deals with 1700 gallons of crude sewage per

FIG. 75.—Electrically driven single-stage pump.

minute against a total head of 170 ft., with an efficiency of 75 per cent.

The belt-driven single-stage pump shown in Fig. 76, when

running at 1750 revolutions per minute, pumps 285 gallons per minute against a total head of 73 ft.

For discharging the cargo of oil tank ships Messrs. Hayward have built several pumps illustrated in Fig. 77.

The pump is a two-stage pump, with the casing split horizontally, allowing the whole rotor to be lifted out, on removal of the top half of the casing.

These pumps were designed to pump 180 tons of oil per hour against a pressure of 70 lb. per sq. in.

FIG. 76.—Belt driven single-stage pump.

Messrs. Hayward, Tyler & Co. have supplied a considerable number of these pumps for boiler feeding. Three of these pumps are at work in the Luton Corporation Electricity Generating Station, two of the pumps each dealing with 8300 gallons of water per hour against a boiler pressure of 190 lb. per sq. in., and one pump for 15000 gallons per hour against 180 lb. per sq. in. These pumps deal with feed water required by the whole plant, and the first have been running continuously for many years.

The high speeds at which a centrifugal pump can be worked make it an ideal machine for direct coupling to a steam turbine, as in Fig. 78.

This single stage pump runs at a speed of 5300 revolutions per minute, when pumping 8000 gallons of feed water per hour against a boiler pressure of 160 lb. per sq. in.

For fire extinction purposes the centrifugal has many ad-

FIG. 77.—Benzine pump.

vantages over the reciprocating type, one being the continuous steady stream that it gives to the jets.

In some cases the ordinary water-supply mains do not give sufficient pressure for fire jets to reach the top of the highest buildings, when the jets are directly connected to the mains by means of hydrants and hose pipes.

In order to get over this difficulty an electrically driven turbo or centrifugal pump is installed on a byepass to the mains.

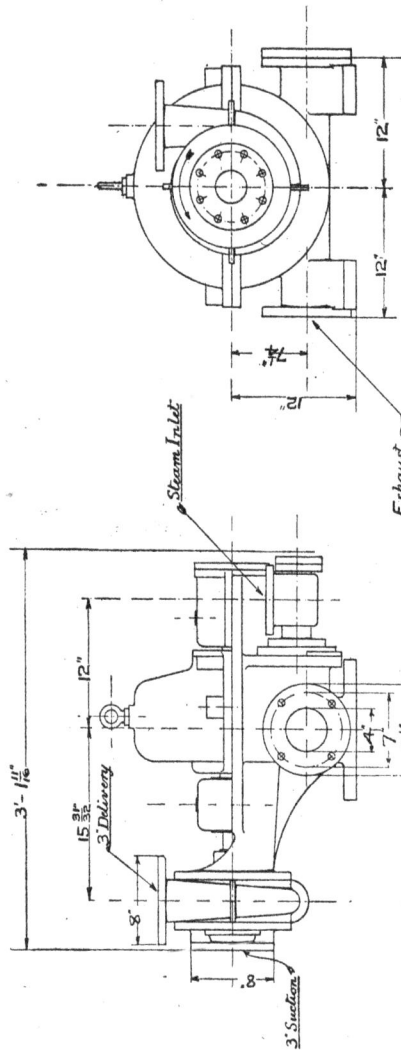

Steam Inlet

Exhaust

12"

12"

7¼"

12"

12"

3' Delivery

15 3/32

8"

.8.

3' Suction

Steam Inlet

4"

7½"

8¼"

3'-1½"

FIG. 78.—Centrifugal boiler feed pump driven by steam turbine.

In the event of a fire, by means of valves dividing the main, the water passes through the centrifugal pump on the byepass, which increases its pressure to desired amount.

A pump installed for this purpose is illustrated in Fig. 79.

This pump is a three-stage pump directly connected to an electric motor which runs at 2200 revolutions per minute.

Fig. 79.—Electrically driven three-stage H.T. turbo-pump.

The pressure at the suction inlet is the usual pressure in the main, namely, 20 lb. per sq. inch.

The increase in pressure when 250 gallons of water per minute passes through the pump is 90 lb. per sq. in., thus giving a pressure of 90 + 20 or 110 lb. per sq. in., in the mains beyond the pump, for supplying the fire streams when required.

CHAPTER XII.

TESTING.

ALL centrifugal or turbine pumps should be carefully tested, in order to see if the pump will give the guaranteed quantity against the required head when the pump is running at the specified number of revolutions.

Another object of testing is to obtain data for use in future designs, to determine the variation in head, quantity, power, efficiency, and speed. The measurements of head, quantity, power and efficiency are plotted as points; the curves drawn through these points represent graphically the performance of the pump, and are called the characteristics of the pump.

The term head is the energy per pound of fluid passing through the pump and is measured by means of the pressure gauges on delivery and the vacuum gauge on the suction.

The pressure gauges should be calibrated by means of a gauge tester.

Fig. 80 shows the arrangement of a pump and the position of the gauges g and g_d, taking the bottom of the suction tank as a datum line.

If v_s and v_d are the velocities at the sections A and B,

Total energy of fluid at A is

$$h_2 + 2 \cdot 3 p_s + \frac{v_s^2}{2g}.$$

Total energy at B is $h_1 + 2 \cdot 3 p_d + \dfrac{v_d^2}{2g}.$

The total energy of the fluid between the sections A and B

$$= 2 \cdot 3 (p_d - p_s) + h_1 - h_2 + \frac{v_d^2 - v_s^2}{2g}.$$

(129)

9

If the diameter of the delivery at B is the same as the diameter of the section at A, then

$$\frac{v_d^2 - v_s^2}{2g} = 0.$$

If h_3 is the difference between the height of the two gauges, that is, $h_1 - h_2$, then the head or energy per pound of fluid between A and B, if the fluid is water, is

$$2\cdot3'(p_d - p_s) + h_3,$$

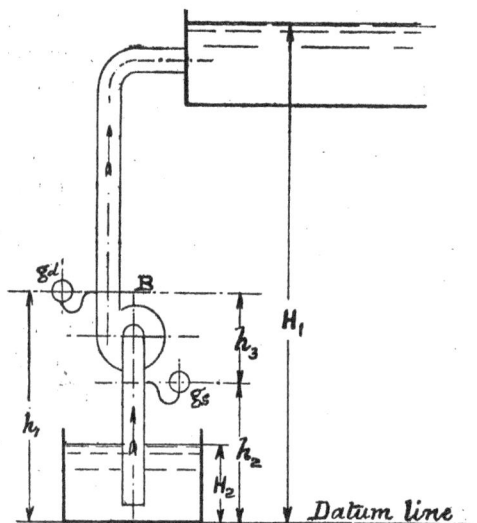

FIG. 80.—Diagram of centrifugal pump installation.

as p_s is frequently measured by a vacuum gauge in inches mercury; and if z be the reading on the vacuum gauge, then the increase of energy per pound of fluid, or head in feet, considering pump runner and casing only, is

$$2\cdot3p_d + 1\cdot129z + h_3.$$

From this, it will be seen that the desired head, or variation in head can be obtained by means of a screw down, or sluice valve on the delivery pipe beyond the section B (see Figs. 81 and 82).

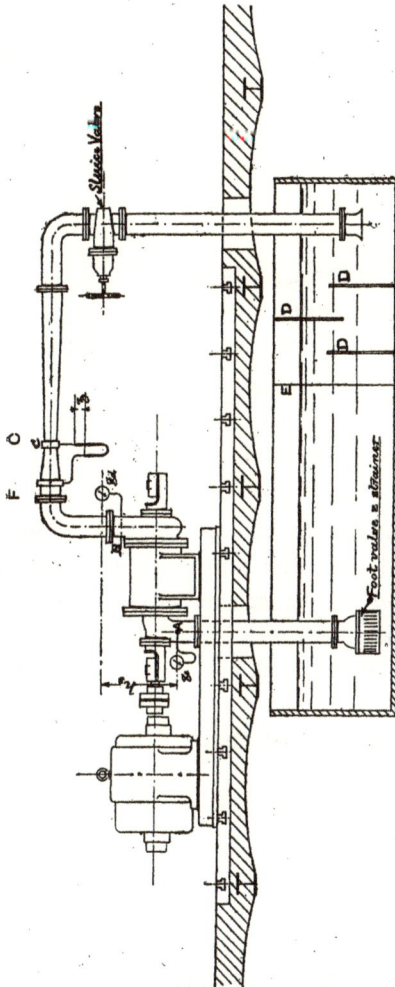

FIG. 81.—Arrangement of pump for testing with venturi meter.

Regarding the quantity of water dealt with, this may be measured by one or more of the following methods :—

(1) Venturi water meter.

(2) Weir tank.

(3) Nozzle in conjunction with pressure gauge.

(4) Nozzle in conjunction with pitot tube.

Fig. 81 shows a venturi meter connected up to a pump; the difference of pressure recorded in the two limbs of the mercury gauge is proportional to the square of the quantity passing through,

or
$$Q = v_2 a_2 \times m \sqrt{z}$$

where z is difference of the heights in the mercury gauge in inches.

a_2 = area of throat at c.

v_2 = velocity of flow through the area a_2.

m = constant for the meter.

If water columns are used the pressures at C and F

then
$$Q = K \frac{a_1 a_2}{\sqrt{a_1^2 - a_2^2}} \sqrt{2g(h_1 - h_2)},$$

h_1 being height of water column at F,

h_2 ,,   ,,   ,,   ,,   ,, C.

There is a maximum and minimum quantity which any venturi meter will pass, therefore if a great variety of pumps are to be tested several venturi meters will be necessary.

The weir tank is shown in Fig. 82 with pump to be tested, and is a very simple method, although it involves calculation of the quantity flowing based upon the dimensions and form of notch or orifice and upon the height of water level above the bottom of the notch.

The notch, V-shaped or it may be rectangular, should be made of brass and bolted to the inner side of the end of tank, making a water-tight joint.

The delivery of the pump is led to end of tank remote from the notch, and two or three baffle plates D and a perforated metal plate E are fixed in the tank to damp out the eddies, thus ensuring smooth flow towards the notch.

The height of water above the bottom of notch is measured by the adjustable hook gauge G, placed as far back from the notch as possible to avoid any error due to velocity of approach.

For a triangular notch containing an angle of 90°

$$Q = 2.635 h^5$$

Q = cubic feet of water flowing per second.

h = height in feet of still water level above bottom corner of notch in feet as measured by the hook gauge.

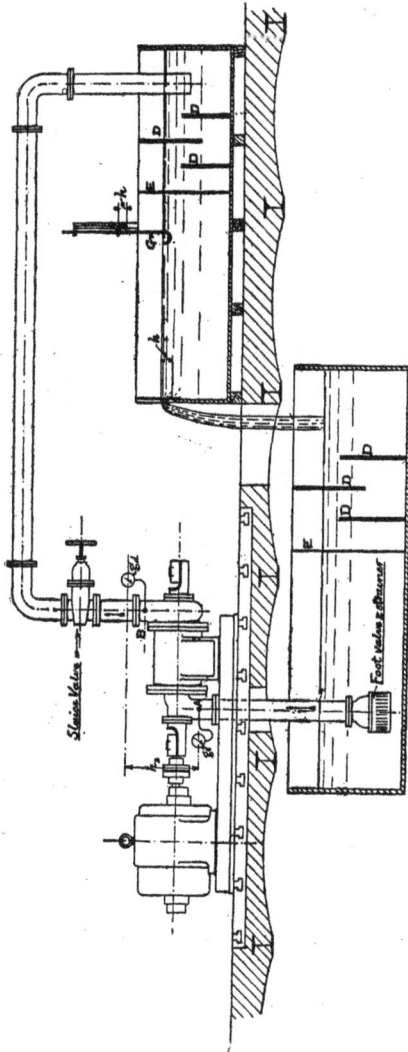

FIG. 82.—Arrangement of pump for testing with wier.

For a rectangular notch, the width b should be $3d$ where d

is the depth of the notch, also the width of the tank channel not less than $(b + 4d)$, and its depth less than $3d$.

$Q = c(b - nh)h^{\frac{3}{2}}$.

$Q =$ cub. ft. per second.

$b =$ width of notch in feet.

$h =$ height of still water level above botton of notch in feet (measured by the hook gauge).

$c = 3\cdot3$.

$n = 0\cdot2$ when b is less than the width of tank.

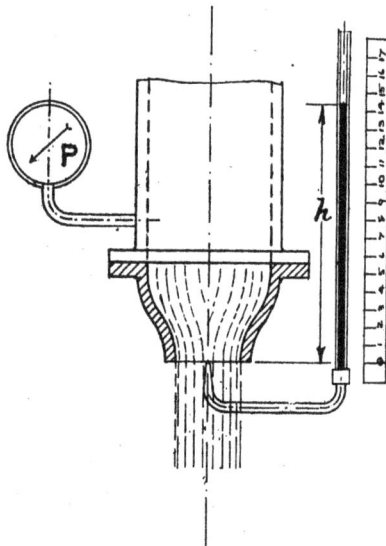

Fig. 83.—Nozzle with gauge and pitot tube.

Then the expression becomes

$Q = 3\cdot3(b - 0\cdot2h)h^{\frac{3}{2}}$.

For the calibrated nozzle with pressure gauge P, Fig. 83,

$$Q = k\sqrt{2g \times 2\cdot3p}$$

where p is the pressure gauge reading in pounds per square inch, k being a constant for the nozzle. This may be put into the form

$$Q = K\sqrt{p},$$

K being a constant containing k, $\sqrt{2g}$ and $\sqrt{2\cdot3}$.

Fig. 83 also shows a nozzle used in conjunction with a pitot tube.

The pitot tube measures the velocity head h at the orifice of the nozzle; the discharge Q is given by

$$Q = Ca\sqrt{2gh}.$$

a = area of nozzle orifice in square feet.

h = height in feet of water in pitot tube.

C = a constant varying from 0·95 to 0·98, which should be found by calibration.

In order to find the efficiency of a pump it is necessary to measure the power given to the pump spindle.

The best method would be to connect some form of transmission dynamometer between pump and the prime mover; this is absolutely necessary if the pump is driven by a turbine or reciprocating engine.

In the case of a pump driven by an engine the engine may be indicated and the over-all efficiency of engine and pump obtained.

For pumps driven by constant speed electric motors, the motor should be tested for all outputs from no load to just over full load by means of some form of brake.

If N_e is the input to motor in electrical horse-power, then

$$N_e = \frac{\text{volts} \times \text{amps.}}{746}.$$

If N_p = brake horse-power corresponding to N_e,

The brake horse-power N_p is the power available at the pump spindle.

The efficiency of the motor $\eta_m = \dfrac{N_p}{N_e}$.

Curves of η_m and N_p can be drawn with N_e as abscissa, enabling N_p to be found from N_e when the motor is coupled up to the pump (see Fig. 84).

The readings taken should be written on a log sheet which may be of the form below :—

Number.	Time.	Revs. per Min. n.	Weir.		Venturi Meter.	Suction.		Delivery.		Total Head $H =$ $h_s + h_d$ h_3 $+\dfrac{v_d{}^2 - v_s{}^2}{2g}$.
			Height h in Feet.	Ga ls. per Min.	Galls. per Min.	Vac. Gauge, ins. of Mercury.	Feet of Water, h_s.	Gauge Reading, lb./sq. ins.	of Water, h_d.	

Water horse-power, $N_w = \dfrac{10 \times \text{Galls /min.} \times H}{33,000}$.	Motor.					Efficiency of Pump $\eta_p = \dfrac{N_w}{N_p}$.	Corrected for Constant Speed n_c.		
	Volts, E.	Current Amperes, C.	$N_e = \dfrac{EC}{746}$	N_p from Curve of Brake Test of Motor.			Total Head, H_c.	Galls. per Min., Q_c.	Power to Drive Pump, N_p.

In testing a pump the valve on delivery should first be throttled so that the pump may give its normal quantity for the given speed ; readings should be taken gradually closing the valve up to no delivery. The valve should then be gradually opened ; readings being taken until the valve is fully open and the pump is giving its maximum quantity.

In testing pumps the speed of the motor varies somewhat with the voltage and the load, the speed being somewhat higher when valve on delivery is closed than when it is fully open.

To correct for speed in order to plot the characteristics for constant speed, the following relations hold for the same efficiency :—

$$H \propto n^2 \text{ when } Q \propto n$$
also
$$N_p \propto n^3 \text{ when } Q \propto n.$$

The reason for this is that the triangles of velocities are similar, and that friction losses are proportional to the square of the velocity, which velocities are proportional to the speed n ;

Therefore $\qquad H_c = H\dfrac{n_c^{\,2}}{n^2}, Q_c = Q\dfrac{n_c}{n},$

$$N_{p_c} = N_p\frac{n_c^{\,3}}{n^3}.$$

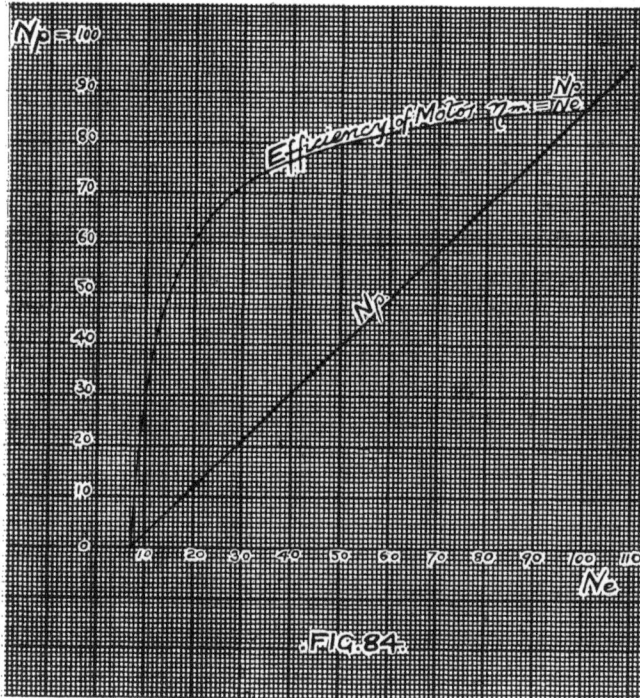

FIG. 84.—Electric motor characteristics.

Suppose that $Q = 600$, $H = 60$, $\eta_p = 0.7$, $N_p = 15.6$ when n is 1440 revolutions, then for a speed $n_c = 1450$ revolutions

$$H_c = 60\frac{1450^2}{1440^2} = 61$$

$$Q_c = 600\frac{1450}{1440} = 605$$

$$N_{p_c} = 15.6\frac{1450^3}{1440^3} = 16.$$

Also, having the characteristics from the test of a pump at say, a speed 1450 revolutions, the characteristics of the same pump may be calculated for any other speed, say 2000 revolutions, from the above relations.

In arranging for testing it is essential that the suction tank should be as large as possible, as a considerable amount of water is required in the circuit.

The delivery water should enter the suction tank at a considerable distance from the entry to the suction, at least 10 ft. or more if quantities over 500 gallons per minute are to be dealt with.

The water where it enters the suction pipe should be as undisturbed as possible; baffles D and a perforated plate E should be fixed between the outlet of the delivery pipe and the inlet to the suction pipe in order to damp out eddies, etc. The depth of water in the suction tank should be about 4 or 5 ft., and it is advisable to have a strainer on the foot valve to keep out any large pieces of foreign substances which very often find their way into the suction tank.

INDEX.

ABERDEEN : THE UNIVERSITY PRESS.